PRACTICAL STATISTICS FOR CHEMICAL RESEARCH

PRACTICAL STATISTICS FOR CHEMICAL RESEARCH

JOHN D. HINCHEN

Monsanto Company, Organic Chemicals Division,
St Louis, Missouri, USA

METHUEN & CO LTD
and
SCIENCE PAPERBACKS

First published 1969
by Methuen & Co., Ltd.,
11 *New Fetter Lane, London E.C.*4
SBN 416 46670 2

First published in Science Paperbacks 1969
SBN 412 20520 3

Set in cold type by
E.W.C. Wilkins & Associates Ltd.
and printed in Great Britain by
Fletcher & Son Ltd., Norwich

Distribution in the U.S.A.
by Barnes & Noble, Inc.

CONTENTS

	Preface	page vi
	Acknowledgements	vii
I	Why statistics?	1
II	Pattern out of chaos	4
III	How to make error behave	7
IV	Is there a difference?	17
V	Is there a relationship?	33
VI	The best-laid plans . . .	48
VII	What does it all mean?	68
VIII	The more things change, the more they are the same	84
	Glossary of terms	91
	Bibliography	100
	Key to use of tables	101
	Tables	104
	Index	115

PREFACE

Churchill Eisenhart's definition of the practical power of a statistical method is the mathematical power of the method times the probability that it will be used. Too often, statisticians become carried away with the pure mathematical beauty of an approach to data analysis, and find that they have lost their audience and with it any hope that the method will ever be of practical value.

In this text, the author hopes to present methods that will be used, to show by word and example how they may be used, and to point out ways of recognizing the proper method to fit a given situation.

Behind the cook-book approach are some key philosophical points which represent some of the later thinking on the application of statistical methods to practical industrial situations.

1. The use of statistical methods should help furnish insight into fundamental mechanisms underlying the behaviour patterns of the data.

2. These methods should assist the experimenter in presenting the data in a clear and readily understood form.

3. The results of these methods should be to provide conclusions which will stand up and which will be readily accepted by others.

4. The simplest approach is often the best in accomplishing these aims.

The methods which will be presented in this book are not new in the statistical field, are not the only methods available, and in many cases may not be the best. They are offered to be used when more precise and more powerful methods may be too costly, too complicated, or not readily accepted on a practical basis.

J.D. Hinchen

ACKNOWLEDGEMENTS

I am indebted to the Literary Executor of the late Sir Ronald A. Fisher, FRS, Cambridge, to Mr Frank Yates, FRS, Rothamsted, and to Messrs Oliver & Boyd, Ltd, for permission to reprint Tables Ia, Ib, Ic, and Id, Table III, and Table VI. These are abridged from Tables V, III, and VI which are published in *Statistical Tables for Biological, Agricultural, and Medical Research* by Fisher and Yates (Oliver & Boyd).

Table II, the Substitute *F* Ratio, is reproduced with permission of the Institute of Mathematical Statistics. This table was originally published in 'On the Ratio of Two Ranges' by F.R. Link, *Annals of Mathematical Statistics* **20** (1949) 257.

Tables IV and V are reproduced with permission of E.S. Pearson from E. Lord's 'The Use of the Range in Place of the Standard Deviation in the *t* Test'(Tables 9 and 10, *Biometrika* **34** (1947) 66).

In Chapters III and IV, tables for estimating standard deviation from the range, and criteria for testing for extreme values are reproduced with permission of W.J. Dixon from *Introduction to Statistical Analysis* by Dixon and Massey (McGraw-Hill), and 'Criteria for Testing Extreme Mean' by Dixon, *Biometrics* **9** (1953).

CHAPTER ONE

WHY STATISTICS?

Everybody is a statistician. This statement may stun some, offend others, and cause still others a great deal of raucous glee. But, in a sense, the statement is true, because everybody at one time or another performs free-of-charge tasks which statisticians get paid fabulous salaries to do.

Everybody makes decisions. These decisions are based on judgement, which in turn is used to interpret information. Most people would like to see their decisions correct at all times, but this is not to be. No one is perfect, and the best we can strive for is to be right as often as possible.

No matter how bull-headed or opinionated an individual may be, no matter how sage or discreet, decisions or judgements are likely to contain two major elements.

1. Evidence or information.
2. A rule of thumb for judgement based on the evidence.

Supplied with these elements, the individual may make a judgement or decision, or alternatively decide that more evidence or information is required.

Anyone that has the problem of driving to work in the morning is faced with the decision of selecting the best route. The data which may influence this decision include:

1. Mileage by each alternate route.
2. Number of traffic lights and stop signs.
3. Road conditions.
4. Traffic.
5. Scenic considerations (animal, vegetable, and mineral).
6. Length of time required for each route.
7. Time of year, weather, etc.

And so on. Some of these factors may depend on others, so that they are not completely independent; and considerable data may need to be taken over a period of time to assure a correct decision.

The rule of thumb to be used by the driver may be:

1. Minimize mileage.

2. Minimize over-all driving time.
3. Minimize wear and tear on driver.
4. Minimize wear and tear on the car.
5. Maximize driving comfort and pleasure.

Whatever rule of thumb is used, he must decide if the routes really differ in respect to this rule of thumb. Here is an example in which all the elements of a statistical analysis are present, yet people make these decisions everyday, without being consciously aware of the data or the rule of thumb.

People move from place to place owing to job transfer, wanderlust, or desire to improve their lot in life. One of the major considerations in the move is the cost of living. In order to make the proper decision, based on the evidence, the man needs to know the cost of such items as food, clothing, housing, transportation, education, etc. His rule of thumb may be to maximize income/outgo, minimize cost of living, improve standard of living at no increased cost, etc. Again, the elements of a statistical analysis are present.

And so it goes – the manager of a baseball team deciding what pitcher to use, who will play left-field, what strategy to use when ahead, etc. – he is trying to increase the probability of winning by making the proper decisions.

In stock market investment, the investor follows trends, cycles, ratio between dividends and interest on savings accounts, etc., to help increase his monetary value. In betting on horses, deciding whether to carry an umbrella, when it is safe to cross the street – all the elements of a statistical decision are present, and all of us, at one time or another, weigh the evidence, apply the rule of thumb, and make the decision. We are all statisticians; so why all the fuss?

Why has such a great mass of methodology been built up in the field of statistics? What is wrong with the intuitive statistical methods used by all of us in making our daily decisions? Why are these not good enough for the decisions we must make about experimental data? The answer lies in two things – in our lack of knowledge of the precise nature of the cause and effect relations present, and in the variability which is present in all data. If we knew precisely all the fundamental chemical and physical laws underlying our data, and if we could rely 100% on the value of each observation, there would be no need for statistics. Everyone would always and everywhere make the right decision, and all people would make the same decision in the same situation.

In the presence of these two gaps in our knowledge, we see different decisions made on the same data by the same person at different times, to say nothing of what happens when different people try to analyse the same data. The science of statistics can bring to this seeming chaos some semblance of order. But perhaps most people (even in industry) may prefer a poor decision arrived at easily to a good one which requires much thought and effort. However, if an individual really wants to make better decisions and improve his chances of being right, the combination of an experienced experimenter and an experienced statistician is hard to beat.

Here are some situations where statistical methods may help.

1. When there are data to be analysed.
2. If a programme for collecting data for analysis is planned.
3. The relative value of certain alternative courses of action is desired.
4. One wants to do better than just flip a coin.
5. Some idea of the future behaviour of a reaction, process, or situation is required.
6. Fundamental insight into basic causes is needed.

Now to the crux of the matter. With all the thousands of volumes on statistical methods, what rule of thumb do we use to decide what method fits a specific case? A thought to keep in mind is that precise experimentation merits precise methodology; decisions on which a large dollar sign is hung warrant comparable expenditures for data collection and analysis. Conversely, the simplest available methods should be used:

1. When experimental control is erratic or random within wide limits.
2. When methods of measurement are relatively imprecise.
3. When screening masses of data from sources requiring some tongue-in-cheek assumptions of validity.
4. In preliminary evaluations aimed at selecting the more promising variables for further study.

In this book, the precise statistical methods and the simple methods covering a given situation will be presented side by side with a minimum of theoretical explanation. 'How to do it' instructions will be included with each, and some specific examples will be discussed.

CHAPTER TWO

PATTERN OUT OF CHAOS

'Alike as two peas in a pod.' Folklore has it that this is the essence of similarity. But did you ever look real closely at two peas from the same pod? They are not always the same size, not necessarily the same shape; one may have minor markings on it which differentiates it from the other, and they may even differ somewhat in colour, gloss, and hardness.

Two ball-bearings fresh from the production line may appear to be identical. Precise measurement and microscopic analysis can turn up differences in diameter, weight, roundness, etc.

Two experiments run in the laboratory under identical conditions do not give the same result; in fact, they might admit two entirely different conclusions as to the course of further experimentation. All these points indicate the existence of variation – the one thing which is present everywhere.

One of the worst things we can do with variation is to sweep it under the rug. It does not do any good to throw away one out of two observations because we think something went wrong with the experiment. Unless we know what went wrong, we may well discard the result which we should have kept. Or perhaps the true answer lies somewhere between (or outside) the two results. No – let us admit the existence of variation (or error, as some call it). Let us bring it out in the open and put it to work for us. When we use the term 'error' we do not mean 'somebody goofed'. It refers to the amount of variation we can expect to be present under the experimental conditions.

What can we expect to cause variation in an experimental situation? Let us look at a few possibilities.

1. *Error of Measurement* An optimistic experimenter sends two samples of a prepared chemical to the laboratory for analysis. Since the two samples were pipetted from the same flask, he expects the results to be the same. They differ by 0·1% of an important component.

A production worker obtains samples from two drums of a water soluble phenolic resin. One has a viscosity of 75 cps, the other has a viscosity of 130 cps.

A series of engine tests are run on lubricants containing equal quantities of a new additive. Piston rating values are 0·7, 3·2, 1·5, 0·9, 2·7.

In all of these familiar situations, differences are obtained where results are expected to be the same. This type of variation is inherent to the method of sampling and testing used. In some cases, it may indicate non-uniformity in the material tested; in others, the sum total of minor differences in weighing, reading, burettes, cleanliness of equipment, evaporation during test, minor differences in temperature, etc., etc.

Many knock-down, drag-out battles between experimenter and tester result when the test variation is considered 'too large'. This topic is not worthy of discussion unless somewhere along the line the basic test error has been sought out, identified, minimized if possible, and published. In Chapters VI and VII, methods for pinning down the sources of test error are discussed.

2. *Known Variables – Controllable* In any experimental situation, certain conditions are held constant during the course of the experiment. A batch of copolymer may be made with a measured quantity of catalyst, at a given reaction temperature, and reacted for a certain definite time. The vessel is free of water, the monomers are charged in a definite proportion, and any other conditions that may affect the result are held fixed. The next portion of the experiment may involve one of these conditions controlled at a new level while the others remain fixed. Proper design of experimental programmes presupposes ability to control the important factors so that the variation due to these factors can be calculated. In Chapter VI, experimental designs for various situations are covered.

3. *Known Variables – Uncontrollable or Controllable Within Limits* A long series of experiments sometimes involves situations where variation due to changes in one or more factors is known to exist, but where these factors cannot be completely controlled. An example is raw-material quality. Different lots of raw material or catalyst used in an experimental programme may sometimes vary in chemical composition, impurities, activity, etc. Since these items may not be within the control of the user and may not easily be worked into a balanced experimental design, the differences should at least be recorded and their effect taken into consideration during the analysis

of the data. Fluctuations in temperature of an oven or bath, when known, should be recorded and the experimental conclusions tempered with this knowledge. Chapter VII gives some methods for dealing with these situations.

4. *Unknown Variables* Every once in a while, a test result, an experimental preparation, yields a value which is 'way out of line' This is our way of saying that the particular result cannot be explained by any of the known variables, or by test and sampling error. Quite often, the tendency is to discard the result, or, with complete intellectual honesty, temporarily suspend consideration of it. However, this type of error or variation may be trying to point out a very important variable which comes and goes at rare intervals, or some test or experimental condition which was not previously considered important. If the error is in a favourable direction, identifying the cause may open up new fields of knowledge and speed up the programme. If the error is in an unfavourable direction, identifying the cause will permit guarding against its repetition in the future. And, if it tends to cloud experimental conclusions, techniques for ruling it out are available (cf. Chapter IV).

Variation is present everywhere. But this is no reason to throw our hands up and say that no conclusions are possible, or, worse yet, to resort to 'black magic' and preconceived notions as to the conclusions which should be drawn from the data. As we will see, some conclusions can always be drawn and the probability of their being correct can be calculated. Every effort should be made to classify error into the four categories listed above. Then we have it where we want it. We can classify, quantify, and modify error, so that instead of clouding the issue it can lead us to more definite conclusions.

›yond group sizes of 10, the estimate becomes relatively poor. ‹arger amounts of data, division of the data at random into ‹ler groups is recommended. The ranges of the smaller groups ‹alculated and averaged, and this average range is then multi- ‹l by a group size factor.

‹n the example given:

Range	=	2·3
Factor	=	0·395
Estimated standard deviation	=	0·91

For the time being, let us use the standard deviation, estimated ‹any of the above techniques, as our measure of experimental ‹or. We all know intuitively that as we run more tests, we get a ‹ter knowledge of the property we are measuring. We also know ‹t the improvement in knowledge depends to some extent on the ‹mber of tests that we run. What is the relationship involved?

Using S to represent the standard deviation, x the value of an ‹dividual observation, $\bar{x}$ the average of n values, the following is ‹und to be true:

$$S_{\bar{x}} = S_x/\sqrt{n}$$

S_x = standard deviation of individual values

$S_{\bar{x}}$ = standard deviation of the average of n values

In the example given, using 0·79 as our estimate of the standard ‹eviation of the individual values, the standard deviation of the ‹verage of 6 results becomes:

$$S_{\bar{x}} = 0{\cdot}79/\sqrt{6} = 0{\cdot}32$$

Up to this point, we have employed the standard deviation as a measure of experimental error, without making any assumptions as to the type of data we have been working with. In many cases, we have previous information or can reason from analogy as to the kind of behaviour we might expect from the data. For example, if we were to continue to run a large number of efflux times on the same sample and to plot the number of results obtained against the value of the result, we would see that the largest number of results is obtained near the middle of the data range. As we get farther and farther away from the middle in either direction, fewer results are obtained. The plot might look something like this:

CHAPTER THREE

HOW TO MAKE ERROR BEHAVE

Vice is a monster of so frightful mien,
As to be hated needs but to be seen;
Yet seen too oft, familiar with her face,
We first endure, then pity, then embrace.

This quotation from Alexander Pope's *Essay on Man* may well be applied to error. In a way, recognition of error in experimentation will lead to its acceptance and eventually to our almost joyful preoccupation with it as a partner in achieving our goals. To avoid misuse of error, for example, ignoring it when we wish to draw a preconceived conclusion or relying on it when we think it can be employed to avoid drawing undesirable conclusions, we first need to understand how it behaves.

A good way to start our investigation of the behaviour of error is to find some way of measuring it. For example, successive observations of efflux time on a viscosity pipette might be as follows (in seconds): 35·3, 37·2, 36·1, 34·9, 35·8, 36·1. This variation represents a form of error and its magnitude can be expressed in several ways. The simplest is the range or spread of the data; the difference between the highest and lowest value. In this case, we have 37·2 to 34·9 or a range of 2·3.

Another measure of variation is obtained by averaging the difference between each value and the over-all average of the data. In this case, absolute values are used.

Observation	*Deviation from Average*
35·3	0·6
37·2	1·3
36·1	0·2
34·9	1·0
35·8	0·1
36·1	0·2
6 | 215·4	6 | 3·4
35·9	0·57

This is called the *mean deviation*, and is often used in analytical chemistry as a measure of the repeatability of a test method.

A third, and perhaps the most important measure of variation is known as the *standard deviation*. This is defined as the root mean square average of the deviations from the mean. In the example given:

Observation	*Deviation from Average*	*Deviation*^{2}
35·3	0·6	0·36
37·2	1·3	1·69
36·1	0·2	0·04
34·9	1·0	1·00
35·8	0·1	0·01
36·1	0·2	0·04

$6\,|\,3{\cdot}14 = 0{\cdot}52$

$\sqrt{0{\cdot}52} = 0{\cdot}72$

For reasons best known to mathematicians, much of the methodology of statistics, and many of the simplified tables and computations have been based on the use of the standard deviation as a measure of variability. Who are we to argue? Later, however, the use of the range as a simplified measure in certain situations will be developed.

In addition, the standard deviation as calculated from a sample may sometimes be used as an estimate of the true standard deviation of a method or process. In these situations, it is found that the standard deviation of a small sample tends to underestimate the true standard deviation. This bias can be compensated for by using one less than the number of observations as divisor of the sum of the squared deviations as given above.

In this example, we would have $\sqrt{(3{\cdot}14/5)} = \sqrt{0{\cdot}628} = 0{\cdot}79$.

Generally, it is easier to calculate the standard deviation of a sample by using an equivalent formula, as follows:

$$\text{Standard deviation} = \sqrt{\left(\frac{\text{sum (obs)}^2 - (\text{sum obs})^2/n}{n-1}\right)}$$

In the example given:

	Observation	*Observa*
	35·3	1246·(
	37·2	1383·8
	36·1	1303·2
	34·9	1218·0
	35·8	1281·6
	36·1	1303·21
Sum	215·4	7736·00

$213{\cdot}4^2 = 46397{\cdot}16$

$$\text{Standard deviation} = \sqrt{\left[\frac{7736{\cdot}00 - \frac{4}{}}{5}\right.}$$

$= \sqrt{0{\cdot}628}$

$= 0{\cdot}79$

This method is preferable to the deviation techn
calculator which can accumulate squares is availab

A method for getting a rapid estimate of the stan
small groups of data makes use of the relationship b
two measures of variation. The range is multiplied b
priate factor and the result is an estimate of the star
(The use of the range is subject to the assumption th
distributed as discussed on p. 11.) This will generall
the calculated standard deviation in a group but, sinc
merely estimates of some true standard deviation, this
expected. The factors for group sizes of 2 to 10 foll
factor is generally close to $1/\sqrt{n}$.

Numbers in Group	*Range Factor*	*1/*
2	0·886	0·7
3	0·591	0·5
4	0·486	0·50
5	0·430	0·44
6	0·395	0·40
7	0·370	0·37
8	0·351	0·352
9	0·337	0·333
10	0·325	0·316

E
For
sma
are
plie

by
er
be
th
n

i
f

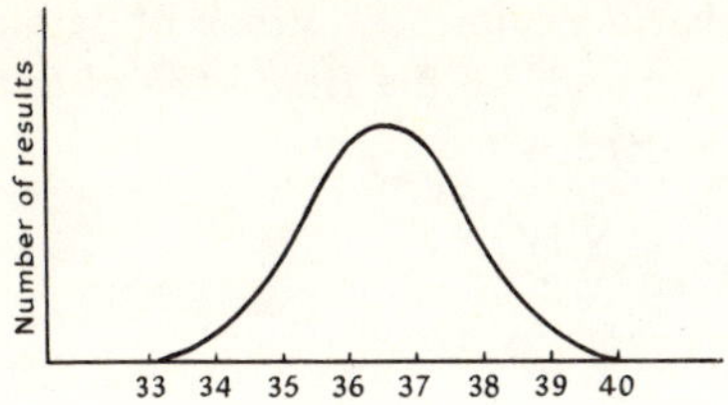

Efflux Time

Most of the conclusions that we will draw in this text are based on the assumption that large quantities of the data we work with would form *distributions* of the same general shape as this. For a thorough discussion of distributions, especially the *normal distribution,* Dixon and Massey [5] or Grant [6] are recommended. Data that conform ideally to the type of curve depicted above will be referred to as normally distributed.

PUTTING ERROR TO WORK

Our estimate of experimental error can be used as a yardstick in two ways: one, to tell us whether one set of data is really different on the average from another; and two, to tell us whether one set of data is more variable than another. Taking the second case first, suppose two test methods are being compared in order to determine whether method B should replace method A. If the variance of B is greater than that of A, the replacement will not be made. A series of determinations is made on a single well-blended sample of material and the results obtained as follows:

Ingredient C (%)

	Method A		*Method B*
	7·05		6·83
	7·00		6·82
	7·05		6·80
	7·01		6·80
	6·85		7·20
	6·96		7·16
	6·94		6·85
	7·13		7·23
	7·00		6·92
	6·95		7·35
Σ =	69·94	Σ =	69·96
Average =	6·99	Average =	7·00

1. Calculate the variance for each group of data. Variance is standard deviation squared, so the final step of taking square root in the calculation is omitted.

$$S_A^2 = \frac{\Sigma A^2 - \frac{(\Sigma A)^2}{n}}{n-1}$$

$$S_A^2 = \frac{489{\cdot}2122 - \frac{4891{\cdot}6036}{10}}{9} = \frac{0{\cdot}0516}{9} = 0{\cdot}0057$$

$$S_B^2 = \frac{489{\cdot}8512 - \frac{4894{\cdot}4016}{10}}{9} = \frac{0{\cdot}4110}{9} = 0{\cdot}0457$$

Where A = the individual result by method A.

2. Determine the ratio of variance B to variance A.

$$F = \text{variance ratio} = 0{\cdot}0457/0{\cdot}0057 = 8{\cdot}02$$

3. Find the critical value of F in the table (Table I).

If we wanted to be 95% certain of being correct in saying that variance B was greater than variance A, we use the 95% F distribution table. For 10, and 10 results, the degrees of freedom (*df*) are 10 – 1 = 9, and 10 – 1 = 9. The critical value for $F_{0{\cdot}95}$ (9,9) is 3·18. The ratio observed is higher. We can say (with a probability of 95% of being correct) that test method B is more variable than test method A.

Tables Ia to Id represent the value of the ratio of the variances which would occur owing to chance if the two variances were in fact from normal distributions which have the same variance. The probabilities given in the tables (0·20 for Ia, 0·05 for Ib, etc.) cover what is called the one-tailed test situation. In using the tables with this probability level, we are interested only in determining whether the F ratio obtained is *larger* than that which could be attributed to chance.

If we are interested in comparing variance A with variance B, and would like to know whether A is larger than B, and also whether A is smaller than B, we use a 'two-tailed' test on the variance ratio. F_1 will be the ratio of variance A to variance B. F_2 will be the ratio of variance B to variance A. Both these ratios are then compared with the critical values in the table. However, the significance level of the table is now doubled, e.g. Table 1b, 0·05 significance level, becomes 0·10 significance level.

For example, if variance A, calculated from 13 points is 10 and variance B, calculated from 25 points is 50,

$$F_1 = \frac{\text{variance A}}{\text{variance B}} = 0{\cdot}2, \text{ with 12 and 24 degrees of freedom.}$$

$$F_2 = \frac{\text{variance B}}{\text{variance A}} = 5{\cdot}0, \text{ with 24 and 12 degrees of freedom.}$$

Comparing the values with the critical values in Table 1b we find that $F_1 < 2{\cdot}2$, but $F_2 > 2{\cdot}5$, so we can say with a 90% chance of being correct (10% chance of being wrong) that the variance of B is greater than the variance of A.

What happens if we use the range instead of the variance as our measure of experimental error?

1. Determine the range of data by each method

$$R_A = 7{\cdot}13 - 6{\cdot}85 = 0{\cdot}28$$
$$R_B = 7{\cdot}35 - 6{\cdot}80 = 0{\cdot}55$$

2. Form the substitute F ratio, R_B/R_A,

$$F_R = 0{\cdot}55/0{\cdot}28 = 1{\cdot}96$$

3. Find the critical value of the substitute F ratio in Table II for 95%. Note – here the actual sample size is used rather than degrees of freedom.

$$F_{R(0{\cdot}95)}\,(10,10) = 1{\cdot}9$$

The value we obtained was 1·96, hence we can draw the same conclusion as above.

This test, as is true of many 'quick and dirty' methods, is not as sensitive for detecting differences as is the more complex F test.

WHAT ARE THE GROUND RULES?

Before we cover the various methods of determining whether differences in level exist between groups of data, a few words on the fundamental assumptions underlying these tests is in order. The concept of statistical inference is expressed very well by Dixon and Massey [5] among others. The logical steps taken in any statistical inference include the following:

1. The assumption is made that the data conform to some basic pattern.
2. A test is made to determine whether this is a logical assumption.
3. The risk that we are willing to run in making either of two types of wrong decision is determined.
   (a) The error of the first kind – that we will say that the data do not conform to the theoretical pattern when in fact they do. The probability level of this kind of error is called α.
   (b) The error of the second kind – that we will say that the data conform to the theoretical pattern when in fact they do not. The probability level of this kind of error is called β.

Some authors have suggested the possibility of the error of the third kind. This error is the probability that we will give the right answer to the wrong question. Although this error is well recognized, there are no statistical tests for guarding against it. Here again, common sense, background knowledge, and statistics must go hand-in-hand.

Many texts have been written going in to great detail on the development and validity of these methods. Some of these references are given in the bibliography. We will proceed with the assumption that the methods discussed here are mathematically valid and have been proved valuable through many years of use.

PROBLEMS

1. What is the estimated standard deviation of the mean $S_{\bar{x}}$ of the following data?

14, 13, 17, 12, 16.

Calculate standard deviation S of the individual observations X.

$$S = \sqrt{\left(\frac{\Sigma x^2 - \frac{(\Sigma x)^2}{n}}{n-1}\right)} = \sqrt{\frac{\left(1054 - \frac{(72)^2}{5}\right)}{4}} = \sqrt{\frac{(1054 - 1036{\cdot}8)}{4}}$$

$$= \sqrt{4{\cdot}30} = 2{\cdot}08$$

$$S_{\bar{x}} = 2{\cdot}08/\sqrt{5} = 0{\cdot}93$$

2. Using the range estimator technique, estimate the standard deviation of the following data obtained as random measurements from a machine operation.

10·00	11·50	8·65	8·88	10·52	10·84
9·68	9·13	10·85	10·13	11·64	9·56
10·94	9·95	10·95	11·28	8·52	8·86
10·48	10·50	10·39	9·53	8·19	8·56
12·31	9·60	8·20	10·02	10·03	10·81

Range of group	1	= 12·31 − 9·68	=	2·63
	2	= 11·50 − 9·13	=	2·37
	3	= 10·95 − 8·20	=	2·75
	4	= 11·28 − 8·88	=	2·40
	5	= 11·64 − 8·19	=	3·45
	6	= 10·84 − 8·56	=	2·29
		Total	=	15·89
		Average	=	2·65

Factor for group size of 5 = 0·430
Estimated standard deviation = (0·430) (2·65) = 1·14

3. Two new sources of raw material were introduced into a process. The process conditions could be adjusted to maintain equal levels of quality, but the batch-to-batch variation might be different for batches made from the two materials. Nine batches of material A were run along with seven batches of material B. The variance of the runs made with material A was 10·72. The variance of the runs made with material B was 3·97. Is there a difference between the two materials in their effect on quality variation, using 90% significance?

This is a two-tailed test situation, since we are interested in variance of A > than variance of B, and variance of B > variance of A.

Form F ratio

$$\frac{\text{Variance A}}{\text{Variance B}} = \frac{10{\cdot}72}{3{\cdot}97} = 2{\cdot}70$$

Critical value from Table Ib for 8 and 6 df = 4·15. The value obtained is less than this, so we cannot say that there is a difference in variance.

If we had formed F ratio

$$\frac{\text{Variance B}}{\text{Variance A}} = \frac{3{\cdot}97}{10{\cdot}72} = 0{\cdot}37$$

we would compare this result to the reciprocal of the critical value in the table for 6 and 8 df. This would be $1/3{\cdot}58 = 0{\cdot}279$. If the observed F were lower than this, we could say that a difference exists and be right 90% of the time. Note – either variance ratio will give the same decision.

CHAPTER FOUR

IS THERE A DIFFERENCE?

One of the least-complicated statistical situations is the one in which two sets of data are being compared to determine whether they differ with respect to some property or variable. The statistical tests governing these comparisons are very simple and differ slightly from one another depending on the individual situation. In this chapter, we shall assume in general that the error in the data is not previously known, but must be estimated from the data.

COMPARISON WITH A STANDARD VALUE

A standard pulp, used in paper testing, gives a tensile strength of 17·0 pounds average when made into paper in a standard formulation and tested in accordance with a definite procedure. A new pulp is purchased and tested over a period of 10 days with the following daily average results: 15·7, 16·2 16·8, 16·2, 15·7, 17·6, 17·1, 16·4, 15·5, 17·0. Does the new pulp show a significant difference in average from the standard?

Procedure	*Calculations*
1. Calculate the average.	1. $\bar{x} = \Sigma x/n = 164{\cdot}2/10$ $= 16{\cdot}42$
2. Calculate the standard deviation. (a) Square values and sum. (b) Subtract (square of sum divided by n). (c) Divide remainder by $(n-1)$. (d) Take square root.	2. $S = \sqrt{\left[\frac{\Sigma x^2 - (\Sigma x)^2}{n-1}\right]}$ $S = \sqrt{\left[\frac{2700{\cdot}48 - 2696{\cdot}16}{9}\right]}$ $S = \sqrt{0{\cdot}48} = 0{\cdot}69$
3. Calculate Student's t. There are several formulas available, each fitting a specific situation. The formula selected is used when an average is being com-	3. $t = \frac{17{\cdot}0 - \bar{x}}{S/\sqrt{n}}$ $t = \frac{17{\cdot}00 - 16.42}{0{\cdot}69/\sqrt{10}}$

pared to some standard value and the data provide the only estimate of the standard deviation.

$$t = 0{\cdot}84(3{\cdot}17) = 2{\cdot}66$$

4. Select α – the probability that we will say a difference exists when in fact it does not.

4. Assume we are willing to run a risk of 5% that we will conclude that a difference exists when in fact the tensile strengths are the same.

5. Compare the calculated value of t with the value in Table III for $(n-1)$ df and the probability level selected.

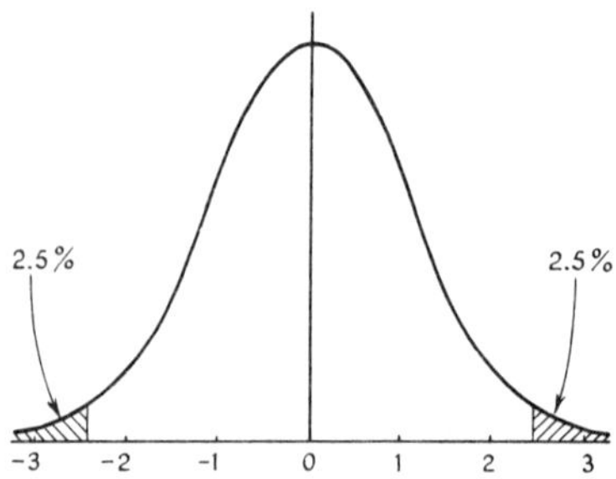

Distribution of t based on random samples of same population $n=10$.

5. Since we are interested in a difference in either direction, we permit t to be either positive or negative and select a t interval which includes 95% of the data. This is found in the column marked $t_{0{\cdot}975}$. The positive value of $t_{0{\cdot}975}$ cuts off 0·025 at the upper end of the scale and 0·025 at the lower. This total corresponds to 0·05. For (10 – 1 =) 9 df, the critical t value is 2·262. Any calculated t value outside the range $\pm 2{\cdot}262$ permits us to state, with a probability of 95% of being correct, that a difference exists. Our calculated value is 2·66, so we say that the new pulp gave significantly different tensile values as compared with the old.

6. Estimate the confidence limits of the new average, based on n data points. For 95% confidence limits, use the critical value of $t_{0{\cdot}975}$ for $(n-1)$ df.

6. Confidence limits are expressed by:

$$\bar{x} + tS/\sqrt{n}$$
$$= 16{\cdot}42 \pm 2{\cdot}262(0{\cdot}69)/\sqrt{10}$$
$$= 16{\cdot}42 \pm 0{\cdot}49$$
$$= 15{\cdot}93 \text{ to } 16{\cdot}91$$

A rapid approximate test which may be substituted for the t test in the situation given proceeds along the same lines.

Procedure	*Calculations*
1. Calculate the average	1. $\bar{x} = \Sigma x/n = 16{\cdot}42/10$ $= 16{\cdot}42$
2. Calculate the range (W).	2. $W =$ highest − lowest $= 17{\cdot}6 - 15{\cdot}5 = 2{\cdot}1$
3. Calculate T_1.	3. $T_1 = (\bar{x} - 17{\cdot}0)/W$ $= (16{\cdot}42 - 17{\cdot}00)/2{\cdot}1$ $T_1 = 0{\cdot}276$
4. Compare with the critical value of T_1 in Table IV for $T_{0{\cdot}975}$ and $n = 10$.	4. The critical value in the table is 0·230. We obtained 0·276, hence we conclude that the new pulp differs from the old.
5. Calculate confidence limits. These will differ slightly from those calculated using t, but should average the same over a large number of trials. This test is about as good as the t test for groups of up to 10 data points and becomes less efficient from there on.	5. Limits $= \bar{x} \pm T_{0{\cdot}975}W$ (95%) $= 16{\cdot}42 \pm 0{\cdot}23(2{\cdot}1)$ $= 16{\cdot}42 \pm 0{\cdot}48$ $= 15{\cdot}94 - 16{\cdot}90$ (95% limits)

COMPARISON OF TWO SETS OF DATA

A more common situation is one in which two sets of data are being compared. For example, in dissolving a phenolic resin in alcohol one might be interested in investigating two different methods of agitation. The variable could be the length of time to dissolve, or the amount dissolved in a given time. Using the latter, we might set up five samples on each of two different shaking devices and measure the percent of sample in solution after 8 hours shaking. Suppose we obtain the following data:

Method A	*Method B*
73·4	73·5
71·1	73·3
72·4	72·7
70·2	72·0
71·7	75·4

Proceed as follows:

Procedure	*Calculations*
1. Calculate the average percentage dissolved for each group.	1. $\bar{x}_1 = 71{\cdot}7$ $\quad \bar{x}_2 = 73{\cdot}4$
2. Calculate the standard deviation for each group	2. $S_1 = 1{\cdot}22$ $S_2 = 1{\cdot}27$
3. Pool the standard deviations. $S_p = \sqrt{\left[\dfrac{(n_1 - 1)S_1^2 + (n_2 - 1)S_2^2}{n_1 + n_2 - 2}\right]}$	3. $S_p = \sqrt{\left[\dfrac{4(1{\cdot}22)^2 + 4(1{\cdot}27)^2}{8}\right]}$ $S_p = 1{\cdot}25$
4. Calculate t for comparison of two means, where S is estimated for each from the data and S_1 does not differ significantly from S_2.	4. $t = \dfrac{\bar{x}_1 - \bar{x}_2}{S_p\sqrt{\left(\dfrac{1}{n_2} + \dfrac{1}{n_2}\right)}}$ $t = \dfrac{71{\cdot}7 - 73{\cdot}4}{1{\cdot}25\sqrt{(0{\cdot}2 + 0{\cdot}2)}}$ $t = \dfrac{-1{\cdot}7}{1{\cdot}25(0{\cdot}64)}$ $= -2{\cdot}13$
5. Compare t with $t_{0{\cdot}975}$ for $(n_1 + n_2 - 2)$ df.	5. $t_{0{\cdot}975} = 2{\cdot}306$ for 8 df $t_{calc} = -2{\cdot}13$ This lies between $\pm 2{\cdot}306$, so the data do not permit us to conclude with 95% confidence that the two methods of agitation differ.
6. Compare t with $t_{0{\cdot}95}$ for	6. $t_{0{\cdot}95} = 1{\cdot}860$. t_{calc} lies

$(n_1 + n_2 - 2)$ df. Strictly speaking, the desired significance level should be specified before seeing the data, otherwise the α and β risks are not as indicated.

outside $\pm 1{\cdot}860$, so we may conclude, with 90% confidence, that a difference exists. If this level of confidence is not sufficient to justify purchase of more expensive equipment, for example, we may wish to obtain more data.

7. Calculate confidence limits for $\bar{x}_1$ and $\bar{x}_2$.
   95% limits for five data points are given by:

   $\bar{x} = S(t_{0\cdot975}$ for 4 df$)/\sqrt{5}$

   90% limits for five data points are given by:

   $\bar{x} \pm S(t_{0\cdot95}$ to 4 df$)/\sqrt{5}$

7. 95% confidence limits for

   $\bar{x}_1 = 71{\cdot}7 \pm 2{\cdot}78(1{\cdot}22)/\sqrt{5}$

   $= 71{\cdot}7 \pm 1{\cdot}52$

   $= 70{\cdot}2$ to $73{\cdot}2$

   95% confidence limits for

   $\bar{x}_2 = 73{\cdot}4 \pm 2{\cdot}78(1{\cdot}27)/\sqrt{5}$

   $= 73{\cdot}4 \pm 1{\cdot}58$

   $= 71{\cdot}8$ to $75{\cdot}0$

   90% limits are:

   for $\bar{x}_1$,

   $71{\cdot}7 \pm 1{\cdot}0 = 70{\cdot}7$ to $72{\cdot}7$

   for x_2,

   $73{\cdot}4 \pm 1{\cdot}1 = 72{\cdot}4$ to $74{\cdot}5$

Another 'quick and dirty' method may be substituted for the above in situations where the standard deviation is not known and the number of data points in each group are the same.

Procedure

1. Calculate $T_d = \dfrac{\bar{x}_1 - \bar{x}_2}{1/2(W_1 + W_2)}$

2. Compare with $P_{0\cdot975}$ for T_d in Table V.

Calculations

1. $T_d = \dfrac{71{\cdot}7 - 73{\cdot}4}{1/2(3{\cdot}2 + 3{\cdot}4)}$

   $T_d = \dfrac{-1{\cdot}7}{3{\cdot}3} = 0{\cdot}515$

2. T_d for $P_{0\cdot975}$ in Table V is $0{\cdot}613$. This is larger than the calculated value, so we cannot say that the groups are different at the 95% significance level.

3. Compare with $P_{0\cdot 95}$ for T_d.

3. T_d for $P_{0\cdot 95}$ is 0·493. The calculated value is larger, hence we may conclude with 90% confidence that the two averages are different.

In both cases above, both the t test and the approximate methods gave the same conclusions.

The procedure for revising α as used here is highly controversial from the purely statistical viewpoint. However, since our purpose is to gain insight rather than be mathematical purists, this approach in the author's opinion may well provide useful information.

Since the data did not provide us with a decision at the 95% level of significance, we have not thereby proved that the two methods of agitation are the same. In fact, we can associate a probability with the statement that they are different as indicated above.

Sometimes we may be interested in knowing how large a difference would be needed in order to reach the desired level of significance. Here concept of least significant difference or LSD* may be of interest.

The LSD is calculated from the formula

$$\text{LSD} = t(S_p)\sqrt{\left(\frac{1}{n_1} + \frac{1}{n_2}\right)}$$

and gives the value of the difference which is just barely significant at the desired probability level. In the example just given,

$$\text{LSD} = 2{\cdot}306(1{\cdot}25)\sqrt{\left(\frac{1}{5} + \frac{1}{5}\right)}$$

$$\text{LSD} = 1{\cdot}84 \text{ for } 95\% \text{ significance}$$

In other words, if the difference between the averages had been 1·84 or greater, we would have concluded that the two groups of data were different at the 95% level of significance.

COMPARISON OF DATA IN PAIRED SETS

Sometimes, a comparison between two sets of data is clouded by the fact that a larger amount of variation exists within each set. This variation may be due to extraneous causes like differences in equipment,

* Not Lysergic Acid Diethylamide

raw materials, day-to-day variation in ambient conditions, etc. It may be that none of this variation is pertinent to the problem at hand, but it may be so large that the LSD for a given level of significance is much larger than the difference one is interested in detecting. In such situations, it is advisable to arrange the comparisons so that individual data points may be compared in pairs, each member of which is obtained under nearly identical levels of the extraneous variables.

An example of how this technique may be beneficial is taken from Paris' *Short Course in Statistical Methods**. The chemist wanted to know if chlorination would affect the abrasion resistance of a particular type of rubber. He selected ten pieces of rubber for the test, and chlorinated one-half of each piece, leaving the other half as it was. Abrasion resistance results were obtained on the twenty pieces.

Test Piece	*Chlorinated*	*Not Chlorinated*	*Difference (d)*
1	12·1	14·7	2·6
2	10·9	14·0	3·1
3	13·1	12·9	−0·2
4	14·5	16·2	1·7
5	9·6	10·2	0·6
6	11·2	12·4	1·2
7	9·8	12·0	2·2
8	13·7	14·8	1·1
9	12·0	11·8	−0·2
10	9·1	9·7	0·6
	$\bar{x}_1 = 11{\cdot}60$	$\bar{x}_2 = 12{\cdot}87$	$\bar{d} = 1{\cdot}27$

To show the advantage of making use of all our knowledge, let us first assume that there were 10 samples chlorinated, 10 not chlorinated, but not identified as to test piece. Using the t test for comparing the two groups as previously illustrated, we obtain:

$$t = \frac{\bar{x}_1 - \bar{x}_2}{S_p \left(\frac{1}{n_1} + \frac{1}{n_2}\right)} \qquad \begin{aligned} S_1 &= 1{\cdot}81 \\ S_2 &= 2{\cdot}07 \\ S_3 &= 1{\cdot}93 \end{aligned}$$

$$t = \frac{11{\cdot}60 - 12{\cdot}87}{1{\cdot}93\sqrt{(0{\cdot}1 + 0{\cdot}1)}} = \frac{1{\cdot}27}{1{\cdot}93\sqrt{0{\cdot}20}} = 1{\cdot}48$$

* Monsanto Organic Research, 1958

The $t_{0\cdot975}$ for $(10 + 10 - 2) = 18$ df is 2·101. Our calculated value is lower, so the difference fails to be significant at the 95% level.

Making use of the fact that the data are logical pairs (i.e. each chlorinated sample corresponds to a particular unchlorinated sample) we proceed as follows:

Procedure	*Calculations*
1. Calculate t for paired samples. With $\bar{d}$ as the average difference (algebraic), S_d the standard deviation of the differences, and n the number of pairs: $t = \bar{d}/S_d/\sqrt{n}$	1. $t = \dfrac{1{\cdot}27}{1{\cdot}13/\sqrt{10}} = 3{\cdot}55$
2. Compare calculated value with $t_{0\cdot975}$ for $(n - 1)$ df.	2. The table value for $t_{0\cdot975}$, $(10 - 1 =)$ 9 df is 2·262. The calculated value of t is higher, so we may conclude, with a 95% chance of being correct, that a real difference exists between the chlorinated and the non-chlorinated samples.

The second t test is more sensitive because it is not affected by the variation in the rubber from sample to sample. The effect of chlorination is more readily detected in the absence of this variation.

This method is good when the extraneous variation is large compared to the difference sought. If the variation within groups is relatively small, pairing reduces the sensitivity of the test.

COMPARISONS WHEN STANDARD DEVIATION IS KNOWN

If sufficient history is available on a test method, a reaction, or a process, so that the experimental error is known, the test for differences between averages of two sets of data becomes more sensitive. In each of the situations considered above, the form of the equation is the same, but it is no longer necessary to calculate the standard deviation. The symbol Z is used instead of t and the critical value is found in the t table for infinite degrees of freedom. This row of the t table corresponds to the table of the normal distribution.

COMPARISONS WHEN THE VARIATIONS DIFFER SIGNIFICANTLY

Using the variance ratio or range ratio tests discussed in Chapter III, we may find that the error or variation in one group of data is significantly greater than that in the other group. In this case, the standard deviations may not be pooled, and a more complex t test results. Also, the degrees of freedom are not simply determined and need to be calculated by the formula given below.

For the t test, comparison of two groups of data, standard deviations found to differ significantly:

$$t = \frac{\bar{x}_2 - \bar{x}_2}{\sqrt{(S_1^2/N_1 + S_2^2/N_2)}}$$

and for the calculated degrees of freedom:

$$\mathrm{df} = \left[\frac{\left(\dfrac{S_1^2}{N_1} + \dfrac{S_2^2}{N_2}\right)^2}{\dfrac{\left(\dfrac{S_1^2}{N_1}\right)^2}{N_1 + 1} + \dfrac{\left(\dfrac{S_2^2}{N_2}\right)^2}{N_2 + 1}}\right] - 2$$

For example, if $S_1^2 = 1 \quad N_1 = 10$

$S_2^2 = 25 \quad N_2 = 20$

$$\mathrm{df} = \left[\frac{\left(\dfrac{1}{10} + \dfrac{25}{20}\right)^2}{\dfrac{\left(\dfrac{1}{10}\right)^2}{11} + \dfrac{\left(\dfrac{25}{20}\right)^2}{21}}\right] - 2$$

$$\mathrm{df} = \left[\frac{\left(\dfrac{27}{20}\right)^2}{\dfrac{\dfrac{1}{100}}{11} + \dfrac{\dfrac{625}{400}}{21}}\right] - 2$$

$$\mathrm{df} = \frac{1{\cdot}82}{0{\cdot}0752} - 2 = 24{\cdot}2 - 2 = 22{\cdot}2$$

Rounding off to the nearest whole number, this gives df = 22 as compared with $N_1 + N_2 - 2 = 28$ when the standard deviations are the same.

Dixon and Massey [5] give a very clear discussion of the various t and Z tests in Chapter IX.

Sometimes a particular value in a set of data is so far 'out of line' that it reduces the sensitivity of the t test when that set is compared with another set. Again, we are indebted to Dixon and Massey [5] for a method. The following excerpt from *Instant Statistics** provides a guide for the proper weeding out of stray values.

The best two out of three

This title does not represent the ground rules for a contest to see who pays for the coffee. It is a well-known expression in research circles when three observations are made on a particular test or experiment and one seems to be widely divergent from the other two. Should the 'wild' value be dropped, and the others averaged, or should the average and spread of all three be reported as factual?

If the wild value is suspect for other reasons, the answer is easy – drop it. For example, if we note our temperature controller went haywire just at the critical moment, or if on looking over our notes we see that we forgot to add the catalyst, the result is considered no good for reasons other than the value itself. But if we have noted nothing suspicious, we should take the approach that all results are innocent (i.e. valid) unless proven guilty (i.e. erroneous).

The attached table gives a good test for determining whether a suspiciously large (or small) value can be reasonably expected to belong to the group with which it is associated. The criterion r, given in the table, can be seen to differ depending on the number of observations in the group. Here is how it works.

Given four GLC results on pentachlorophenol of 88·2, 86·7, 87·9, and 87·7%, obtained under comparable conditions, are we justified in dropping the 86·7% result? *Note* – let us assume we want a 95% chance of being correct.

$$r = \frac{86{\cdot}7 - 87{\cdot}7}{86{\cdot}7 - 88{\cdot}2} = \frac{-1{\cdot}0}{-1{\cdot}5} = 0{\cdot}667$$

* Monsanto Organic Research, 1962

Note in the table, the critical value at 0·05 (95% significance level) for four observations is 0·765. The test value is lower than this, so we cannot reject the suspected observation at the desired level of significance.

Note the equation changes form for larger groups of data. To use the table properly, we must define the following:

x_n = the suspected value

x_{n-1} = the nearest to it

x_{n-2} = the nearest, once removed

x_1 = the furthest away from the suspected value

x_2 = the next furthest away

etc.

Critical values and criteria for testing for extreme values

α	0·30	0·20	0·10	0·05	0·02	0·01	0·005	*Criterion*
$100(1-\alpha)$	70	80	90	95	98	99	99·5	
n = 3	0·684	0·781	0·886	0·941	0·976	0·988	0·994	
4	0·471	0·560	0·679	0·765	0·846	0·880	0·926	
5	0·373	0·451	0·557	0·642	0·729	0·780	0·821	$r_{10} = \frac{x_n - x_{n-1}}{x_n - x_1}$
6	0·318	0·386	0·482	0·560	0·644	0·698	0·740	
7	0·281	0·344	0·434	0·507	0·586	0·637	0·680	
8	0·318	0·385	0·479	0·554	0·631	0·683	0·725	
9	0·288	0·352	0·441	0·512	0·587	0·635	0·677	$r_{11} = \frac{x_n - x_{n-1}}{x_n - x_2}$
10	0·265	0·325	0·409	0·477	0·551	0·597	0·639	
11	0·391	0·442	0·517	0·576	0·638	0·679	0·713	
12	0·370	0·419	0·490	0·546	0·605	0·642	0·675	$r_{21} = \frac{x_n - x_{n-2}}{x_n - x_2}$
13	0·351	0·399	0·467	0·521	0·578	0·615	0·649	
14	0·370	0·421	0·492	0·546	0·602	0·641	0·674	
15	0·353	0·402	0·472	0·525	0·579	0·616	0·647	
16	0·338	0·386	0·454	0·507	0·559	0·595	0·624	
17	0·325	0·373	0·438	σ·490	0·542	0·577	0·605	
18	0·314	0·361	0·424	0·475	0·527	0·561	0·589	
19	0·304	0·350	0·412	0·462	0·514	0·547	0·575	$r_{22} = \frac{x_n - x_{n-2}}{x_n - x_3}$
20	0·295	0·340	0·401	0·450	0·502	0·535	0·562	
21	0·287	0·331	0·391	0·440	0·491	0·524	0·551	
22	0·280	0·323	0·382	0·430	0·481	0·514	0·541	
23	0·274	0·316	0·374	0·421	0·472	0·505	0·532	
24	0·268	0·310	0·367	0·413	0·464	0·497	0·524	
25	0·262	0·304	0·360	0·406	0·457	0·489	0·516	

PROBLEMS

1. A process was run under two sets of operating conditions. The yield values for condition A and condition B are given below. Is there reason to believe, at the 0·05 level of significance, that these conditions result in different yields?

Method: two-tailed t test. First determine whether the variations are compatible.

	A	B
	97·2	96·5
	96·5	96·9
	97·4	96·3
	96·8	96·0
	96·8	96·1
	484·7	96·2
		578·0
Average	96·94	96·33
Sum of squares =	46987·33	55861·20

$$S_A^2 = \frac{46987{\cdot}33 - \dfrac{(484{\cdot}7)^2}{5}}{4} = 0{\cdot}1275$$

$$S_B^2 = \frac{55861{\cdot}20 - \dfrac{(578{\cdot}0)^2}{6}}{5} = 0{\cdot}1080$$

$$F = \frac{0{\cdot}1275}{0{\cdot}1080} = 1{\cdot}18$$

$$F_{0{\cdot}02}\ \text{(two-tailed, Table Ic)}_{4,5} = 11{\cdot}4$$

Variances may be considered compatible.

$$\text{Pooled variance} = S_p^2 = \frac{(0{\cdot}1275)\,4 + (0{\cdot}1080)\,5}{9} = 0{\cdot}1167$$

$$t = \frac{\bar{x}_A - \bar{x}_B}{S_p\sqrt{\left(\dfrac{1}{n_1} + \dfrac{1}{n_2}\right)}} = \frac{96{\cdot}94 - 96{\cdot}33}{0{\cdot}34\sqrt{\left(\dfrac{1}{5} + \dfrac{1}{6}\right)}} = \frac{0{\cdot}61}{0{\cdot}34\sqrt{0{\cdot}37}} = 2{\cdot}90$$

$$\text{df} = n_1 + n_2 - 2 = 9 \qquad t_{0{\cdot}05}\ 9\,\text{df} = 2{\cdot}262$$

If we make the statement that the differences are real, we have a 95% chance of being correct.

2. A method of preparing plastic samples for testing is advocated with the claim that this method will result in higher average tensile strength values. Ten samples are prepared by the new method and ten by the old. The results are given below. Is there reason to believe, at the 0·05 level of significance that the new method gives higher results than the old. (Use one-tailed t test.)

	New Method	*Old Method*
	2600	2500
	2500	2550
	2600	2500
	2650	2600
	2550	2650
	2600	2500
	2550	2550
	2650	2500
	2600	2550
	2600	2600
	25900	25500
Average	2590	2550
	$S_n^2 = 2111$	$S_o^2 = 2778$

$$\text{Variance ratio } F = 2778/2111 = 1{\cdot}31$$

For 9 and 9 degrees of freedom, this number is much lower than any value in Tables Ia to Id. It is all right to pool the variances:

$$S_p = 49{\cdot}4$$

$$t = \frac{2590 - 2550}{49{\cdot}4\sqrt{\left(\frac{1}{10}+\frac{1}{10}\right)}} = \frac{40}{49{\cdot}4(0{\cdot}45)} = 1{\cdot}80$$

$t_{0{\cdot}05}$ (one-tailed) 18 df = 1·734 Significance 95%

3. (a) Three observations were obtained on the formaldehyde content of a formalin sample. The results were:

32·3, 32·7, 31·0

Are we justified at $\alpha = 0{\cdot}05$ in discarding the third result as a stray or wild value?

$$r_{10} = \frac{31{\cdot}0 - 32{\cdot}3}{31{\cdot}0 - 32{\cdot}7} = 0{\cdot}765 \qquad \text{Critical value} = 0{\cdot}941$$

Difference not significant, no reason to discard result.

(b) Additional values were obtained on the same sample. All results are tabulated below. Is there now sufficient evidence ($\alpha = 0{\cdot}05$) for discarding the third result?

32·3	31·9	32·3	32·6	31·9
32·7	32·1	32·2	32·0	32·1
31·0	32·4	32·1	32·5	

$$n = 14 \qquad r_{22} = \frac{31{\cdot}0 - 31{\cdot}9}{31{\cdot}0 - 32{\cdot}5} = 0{\cdot}600$$

$$\left.\begin{aligned} x_n &= 31{\cdot}0 \\ x_{n-2} &= 31{\cdot}9 \\ x_3 &= 32{\cdot}6 \end{aligned}\right\} \quad r_{22(14)} = 0{\cdot}546$$

The calculated value of r_{22} exceeds the value in the table. We are justified in discarding the third result.

INSTANT STATISTICS

Precision, like virtue, is its own reward

The age of chivalry may be dead, but the age of craftsmanship is still with us. Any artisan worth his salt – whether he be a lathe operator turning out metal pieces, a laboratory technician performing chemical analyses, an analytical or physical chemist developing methods for others to use – is interested in improving and perfecting his work. But some of our people are so skilled, some of their methods so good, that it is sometimes hard to tell whether an improvement has been made or not.

One rule of thumb for measuring improvement in a test method, for example, is how well an analysis on a given sample will check when repeated. This estimate of precision, known as repeatability, can be used for comparing two methods when run on the same sample by the same individual using the same equipment. Another such rule of thumb is how well a test method can repeat itself when tests on the same sample are run by different analysts, on different equipment over a period of time. This criterion, known as reproducibility, is also a critical characteristic of a good analytical method.

In evaluating two test methods for repeatability or reproducibility, the variances (standard deviation squared) of the two can be compared by means of the variance ratio or F test. This is a tried and true statistical technique which has many other uses, as well as being a critical part of analysis of variance.

Statisticians (some of whom are craftsmen) like to improve their methodology. Some statisticians feel that a simpler statistical method than the one normally used constitutes an improvement. 'Quick and dirty' methods result; one of which can be used as a substitute for the F ratio test.

Two test methods run on the same sample give the following sets of results:

Method I – 47·2, 50·4, 51·3, 50·8, 47·9, 48·6, 50·6, 47·0, 51·2, 48·1.

Method II – 49·9, 48·9, 49·6, 50·1, 49·2, 50·6, 49·6, 49·4, 48·6, 50·1

To decide quickly, and with a 95% chance of being correct, as to whether method II is actually more precise than method I, perform the following simple procedure:

1. Determine the range of each method (difference between the highest and lowest value).

$$R_1 = 51{\cdot}3 - 47{\cdot}0 = 4{\cdot}3$$
$$R_2 = 50{\cdot}6 - 48{\cdot}6 = 2{\cdot}0$$

2. Form the ratio of the larger to the smaller range.

$$R_1/R_2 = 4{\cdot}3/2{\cdot}0 = 2{\cdot}15$$

3. Compare this value to the table for sample size 10 in the numerator and 10 in the denominator. The critical value in the table is 1·9.

4. Make the statement, with a 95% chance of being correct, that method I is less precise (or reproducible) than method II.

Try this on your own data, where you are interested in reducing variation of a test, a process or what have you. Here is a table for combinations of samples from 2 to 10. *Note*: sample size means number of tests run, not grams or millilitres used.

Sample Size for Numerator

Sample Size for Denominator	2	3	4	5	6	7	8	9	10
2	12·7	19·1	25·0	28·0	29·0	31·0	32·0	34·0	36·0
3	3·19	4·4	5·0	5·7	6·2	6·6	6·9	7·2	7·4
4	2·02	2·7	3·1	3·4	3·6	3·8	4·0	4·2	4·4
5	1·61	2·1	2·4	2·6	2·8	2·9	3·0	3·1	3·2
6	1·36	1·8	2·0	2·2	2·3	2·4	2·5	2·6	2·7
7	1·26	1·6	1·8	1·9	2·0	2·1	2·2	2·3	2·4
8	1·17	1·4	1·6	1·8	1·9	1·9	2·0	2·1	2·1
9	1·10	1·3	1·5	1·6	1·7	1·8	1·9	1·9	2·0
10	1·05	1·3	1·4	1·5	1·6	1·7	1·8	1·8	1·9

CHAPTER FIVE

IS THERE A RELATIONSHIP?

One way of gaining insight into the behaviour of a measurable variable under study is to see how it reacts when some other variable changes. If one variable changes when the other does, whether in the same or opposite directions, the variables are said to be related. The statistical techniques used to study such relationships are classified as correlation or regression techniques. Correlation problems are those in which the two variables under study are free to vary at random. Regression problems are those in which one variable is allowed to vary at random, at several fixed values of the other. The methods used for analysing the data are essentially the same. See Dixon and Massey [5] for more detailed discussion of the differences between regression and correlation.

The first step we usually take in analysing this kind of data is to plot one variable against the other on a square-ruled chart. Let us look at the following data, for example.

Wet Tensile Strength (psi)	*Sizing Added* (%)
1·0	0·5
1·2	0·7
1·1	0·7
1·3	0·7
1·2	0·8
1·5	0·8
1·6	0·9
1·5	0·9
1·6	1·0
1·9	1·3
1·9	1·4
2·0	1·5
1·8	1·5

We plot the wet tensile strength on the vertical axis, and the percent sizing on the horizontal axis.

With data as obviously related as this, we can almost draw a line by eye through the points and say that we have established the relationship. With less obviously related data, we have a problem.

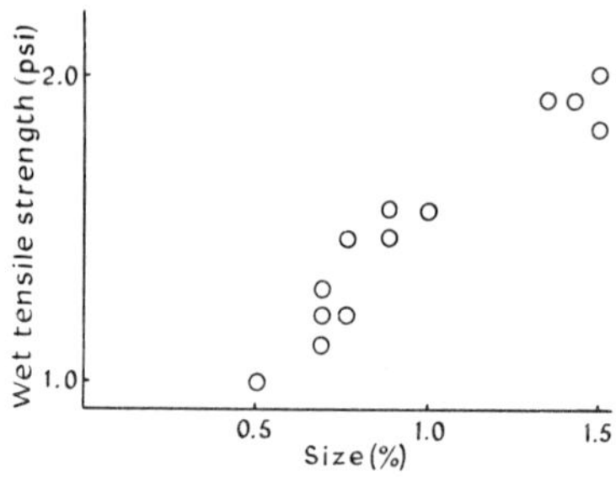

Suppose someone asks:

1. How sure are you that there is a relationship?
2. What value of wet tensile can we expect to get at 1·2% size?
3. How much spread in wet tensile can we expect at a given size concentration?

In a problem of this kind, correlation analysis can be used. We proceed as follows:

Procedure	*Calculations*
1. Set up the data in tabular form, pairing the corresponding x and y values as shown	1.

x	y
0·5	1·0
0·7	1·2
0·7	1·1
0·7	1·3
0·8	1·2
0·8	1·5
0·9	1·6
0·9	1·5
1·0	1·6
1·3	1·9
1·4	1·9
1·5	2·0
1·5	1·8

2. Calculate Σx (i.e. the total of all the x values).	2. Σx = 12·7
3. Calculate Σy (i.e. the total of all the y values).	3. Σy = 19·6

4. Find Σx^2 (i.e. square each x value and add).

4. $\Sigma x^2 = 13{\cdot}77$

5. Find Σy^2 (i.e. square each y value and add).

5. $\Sigma y^2 = 30{\cdot}86$

6. Find Σxy (i.e. multiply each x value by its corresponding y value and add these products).

6. $\Sigma xy = 20{\cdot}40$

7. Square the total of the x values and divide by the number of pairs. This is $(\Sigma x)^2/n$.

7. $(\Sigma x)^2/n = (12{\cdot}7)^2/13 = 12{\cdot}41$

8. Square the total of the y values and divide by the number of pairs. This is $(\Sigma y)^2/n$.

8. $(\Sigma y)^2/n = (19{\cdot}6)^2/13 = 29{\cdot}55$

9. Multiply the total of the x values by the total of the y values and divide by n. This is $\Sigma x \Sigma y/n$.

9. $\Sigma x\Sigma y/n = 19{\cdot}15$

10. Solve for the correlation coefficient r.

$$r = \frac{\Sigma xy - (\Sigma x\Sigma y/n)}{\sqrt{\left[\left(\Sigma x^2 - \frac{(\Sigma x)^2}{n}\right)\left(\Sigma y^2 - \frac{(\Sigma y)^2}{n}\right)\right]}}$$

or in terms of the steps in this procedure (procedure number given in parentheses)

$$r = \frac{(6) - (9)}{\sqrt{[(4) - (7)][(5) - (8)]}}$$

10. $$r = \frac{20{\cdot}40 - 19{\cdot}15}{\sqrt{[(13{\cdot}77 - 12{\cdot}41)\ (30{\cdot}86 - 29{\cdot}55)]}}$$

$$r = \frac{1{\cdot}25}{\sqrt{[(1{\cdot}36)\ (1{\cdot}31)]}} = \frac{1{\cdot}25}{1{\cdot}34} = 0{\cdot}933$$

11. Compare r with the value in the Table VI for $(n-2)$ df. Use the column representing the risk you are willing to run saying a relationship exists when in fact it does not.

11. In Table VI, $r_{0{\cdot}05} = 0{\cdot}553$ for 11 df, $r_{0{\cdot}01} = 0{\cdot}684$ and $r_{0{\cdot}001} = 0{\cdot}801$. The calculated value of r is greater than any of these, so we can say with 99·9% confidence that a relationship exists.

Once the existence of the relationship has been established, the calculations already made can be used to fit a straight line to the data. The procedure used is as follows:

Procedure	*Calculations*
1. Calculate the average of the x values and the average of the y values. These are $\bar{x} = \Sigma x/n \quad \bar{y} = \Sigma y/n$	1. $\bar{x} = 12{\cdot}7/13 = 0{\cdot}98$ $\bar{y} = 19{\cdot}6/13 = 1{\cdot}51$
2. Calculate the slope, b, of the line. $b = \dfrac{\Sigma xy - \dfrac{\Sigma x \Sigma y}{n}}{\Sigma x^2 - \dfrac{(\Sigma x)^2}{n}}$	2. $b = 1{\cdot}25/1{\cdot}36 = 0{\cdot}92$
3. Substitute these results in the equation $y = \bar{y} + b(x - \bar{x})$. This equation may be used to draw a straight line through the data points plotted on the chart. For the data given, the line is shown below.	3. $y = 1{\cdot}51 + 0{\cdot}92(x - 0{\cdot}98)$ $y = 0{\cdot}92x + 0{\cdot}61$ This is the equation for estimating y at a given value of x. At a size concentration of 2·0%, we would expect to obtain a wet tensile of $y = 0{\cdot}92(2{\cdot}0) + 0{\cdot}61$ $= 1{\cdot}84 + 0{\cdot}61$ $= 2{\cdot}45.$

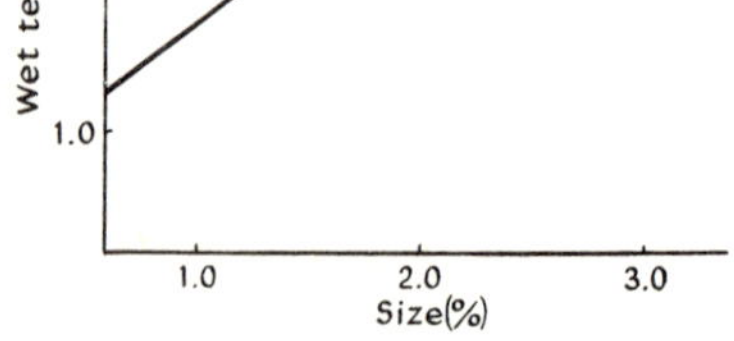

Since error in measurement exists, we would expect that tests run at various size concentrations would not give wet tensile results exactly on the line. From the information already obtained we can estimate the 95% limits of the spread of the data around the line, and use these to predict the range of future results..

Procedure	*Calculations*
1. Calculate the standard deviation of all the wet tensile results. $$S_y = \sqrt{\left[\frac{\Sigma y^2 - \frac{(\Sigma y)^2}{n}}{n-1}\right]}$$	1. $S_y = \sqrt{(1{\cdot}31/12)} = \sqrt{0{\cdot}109}$ $= 0{\cdot}33$
2. Multiply this by $\sqrt{(1 - r^2)}$. This is $S_{y.x}$, the standard deviation of y at a given value of x.	2. $S_{y.x} = 0{\cdot}33\sqrt{[1 - (0{\cdot}933)^2]}$ $= 0{\cdot}33\sqrt{0{\cdot}13} = 0{\cdot}12$
3. The 95% limits are given by $y + t(S_{y.x})$, where y is the value predicted from the equation, and t is the value in the t table at 0·025 for $(n - 2 =)$ 11 df.	3. 95% limits are $y \pm 2{\cdot}20(0{\cdot}12) = 0{\cdot}26$
4. Construct two lines parallel to the calculated line, at a distance of $t(0{\cdot}12)$ above and below the line. Future observations are expected to fall between these lines 95% of the time, near the centre of the data, < 95% away from the centre. Plot the actual values to compare with the limits. *Note*: to be statistically correct, these lines should be adjusted to bend away from the calculated line in proportion to the distance from $\overline{x}$.	4. 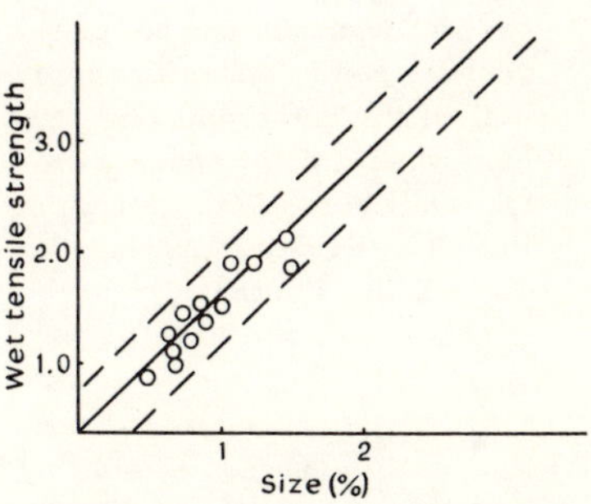

In many actual situations, a search for the most important variables may consist in testing correlations of these variables with some desired characteristic of a product. We may have a large number of these variables to test, and may find that the use of the method just described entails more work than it is feasible to handle. A rapid approximate method is available, which may help to screen out the most important variables. This method is known variously as 'Quadrant Sum Test' or 'Rapid Corner Test for Association'.

Procedure	*Calculations*
1. Plot the data on square ruled coordinate paper, one variable on each axis.	It goes something like this:
2. Divide the data into four quadrants by drawing a vertical line through the median (middle point) of the x values and a horizontal line through the median of the y values.	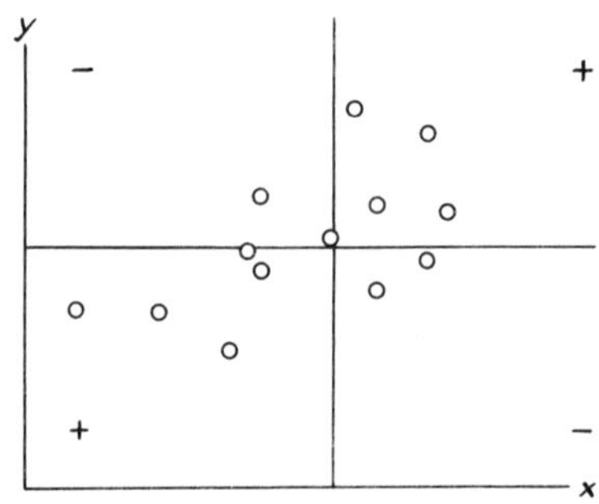
3. Label the north-east and south-west quadrants plus, the north-west and south-east quadrants minus.	
4. Starting at the right side of the chart, lay a straight edge on the paper, parallel to the vertical axis. Moving this toward the centre, count the number of successive points appearing on one side of the horizontal line, stopping when a point appears on the other side of the horizontal line 1 . Give this number the sign of the quadrant.	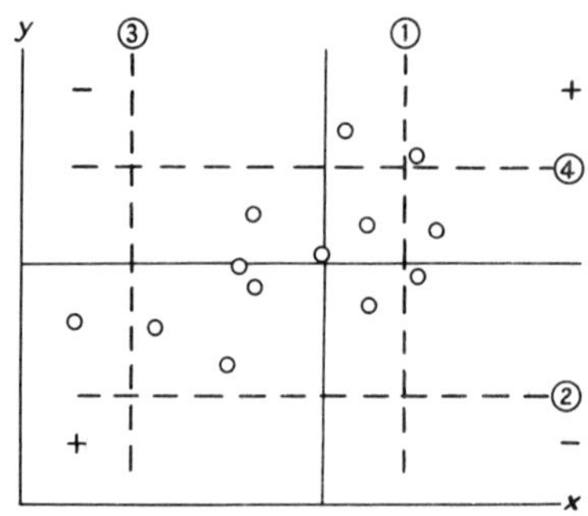
5. Repeat this procedure, starting from the bottom using a horizontal straight edge, 2 then from the left, using a vertical straight edge 3 , and again from the top, using a horizontal straight edge 4 .	1. First count, right to left = + 2 2. Second count, bottom upward = + 3 3. Third count, left to right = + 3 4. Fourth count, top downward = + 2 Quadrant sum = +10
6. Add these counts.	6. Quadrant sum = +10

7. Compare this total to the following table.

7. The total is +10. This is larger than the table figure.

Significance Level	*Quadrant Sum*
90%	± 9
95%	±11
98%	±13
99%	±14

Olmstead and Tukey [7]

For 90% significance, and smaller than the 95% figure. We say with a probability of 90 – 95% that a relationship exists.

In addition to the 'quick and dirty' method used above to see whether a relationship exists, there are graphical methods which may be used to fit the best straight line to a set of data. There are several methods which improve your ability to draw a line which describes a relationship, without extensive calculation. One of the better of these is described below.

Procedure

1. Plot the data on a 'scatter diagram'.

2. Divide the data into 3 groups by means of vertical lines as shown, so that each group contains the same number of points. Put any extra points into the middle group.

3. Find the median point of all the data 1 (1/2 points above and below the horizontal median, 1/2 points right and left of the vertical median).

4. Find the median point of each outside group 2 and 3 .

5. Lay straight edge between the medians of the outside group, and move it (holding it parallel

Calculations

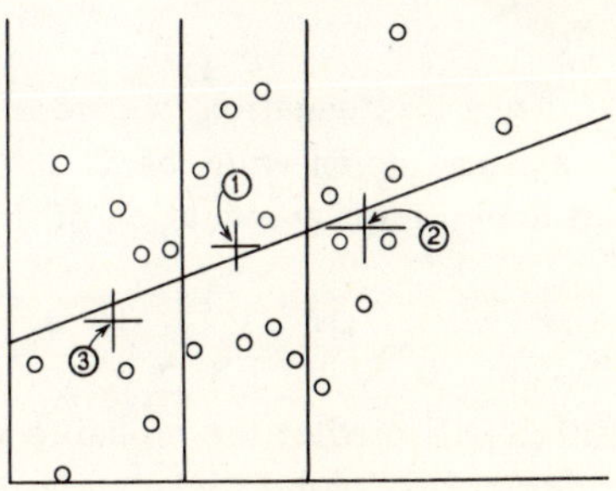

1 = overall median

2 = median of right-hand group

3 = median of left-hand group

Quenouille [8]

to its original position), 1/3 of the distance to the overall median.

6. Draw the line in this position.

In the example given, the line coincides exactly with the least-squares line calculated for estimating y from x.

Sometimes, the relationship between two variables is obscured by the presence of a third variable which may or may not be related to the other two. In order to clarify the relationship, we may want to remove or account for the effect of the third variable. For example, we may want to find the effect of plasticizer viscosity on the hardness of a resin-plasticizer mix. But some plasticizers are more efficient than others, and this factor may obscure the relationship. What we are really looking for is some way of separating out these effects. For example, in the simpler linear correlation study, we came up with an equation of the form

$$y = mx + b$$

From this equation, we were able to predict the probable value of y at some given value of x. In the plasticizer viscosity example, we are looking for an equation of the form

$$y = m_1x_1 + m_2x_2 + b$$

which will predict the probable value of y at some combination of values of x_1 and x_2.

The least-squares method is described in Brownlee [1]. This treatment affords an excellent cookbook approach to problems of this nature and is recommended.

A graphical method can be applied to problems of this nature and has the same basic assumptions as the least-squares method, namely that the x variables do not interact with one another.

Given the following data, find a linear equation of the form

$$y = c_0 + c_1x_1 + c_2x_2$$

which fits the data. (cf. Brownlee for least-squares approach.)

y	x_1	x_2
− 6·6	1	5
12·4	3	2
0·2	6	9
20·8	9	7
31·1	9	3
5·8	2	1
1·3	5	7
27·2	9	6
19·3	8	7
4·2	2	3
2·7	3	5
− 17·1	1	9
6·7	8	9
− 1·4	2	5
12·1	7	7
22·8	7	4
26·0	8	3
24·2	6	1
3·9	1	1
24·0	5	1

Procedure — *Calculations*

1. Plot y against each variable and select the 'best' plot by use of the quadrant sum test. Assuming y vs x_1 is best in this case.

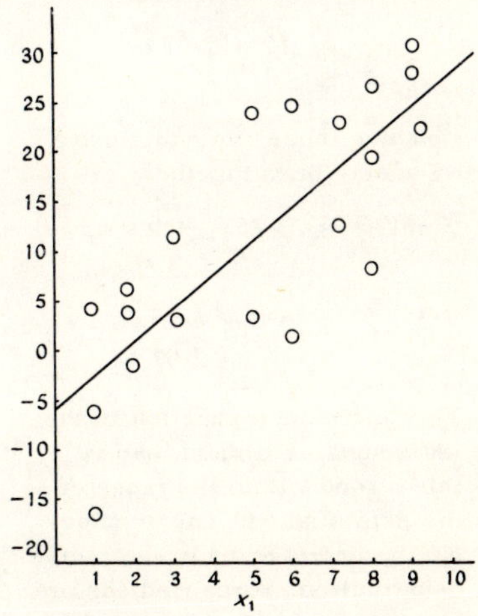

2. Fit a straight line to the data by the graphical method previously described.

3. Determine the ± deviations from the line and plot these against the corresponding value of x_2. *Note*: If actual is above line, deviation is positive, if below line, deviation is negative.

4. Draw the line for the deviations vs x_2.

5. Construct symmetrical limits which include 95% of the data.

6. Obtain the equation of the first line from the points $y_0 = -6$, $x_0 = 0$ and $y' = 25$, $x' = 9$ as follows:

$$\text{Pred } y = y_0 + \left(\frac{y' - y_0}{x' - x_0}\right)x_1$$

$$\text{Pred } y = -6{\cdot}0 + \left(\frac{31}{9}\right)x_1$$

$$= 3{\cdot}45x_1 - 6{\cdot}0$$

7. Obtain the equation of the second line from the points $d_0 = 12$, $x_0 = 0$ and $d' = -12$ $x' = 9$.

$$\text{Pred } d_0 = d_0 + \left(\frac{d' - d_0}{x' - x_0}\right)x_2$$

$$\text{Pred } d_0 = 12 + \left(\frac{-24}{9}\right)x_2$$

$$= 12 - 2{\cdot}67x_2$$

8. Combine these two equations by adding them together.

$$\text{Final } y = 3{\cdot}45x_1 - 6{\cdot}0 + 12{\cdot}0 - 2{\cdot}67x_2$$

$$\text{Final } y = 6{\cdot}0 + 3{\cdot}45x_1 - 2{\cdot}67x_2$$

This is an approximation to the least-squares method, and is fairly good within the range of the data studied. The accuracy can be improved by more precise construction. Since medians are used instead of least squares for fitting the line, some bias is expected, which will increase with the amount of error in the data.

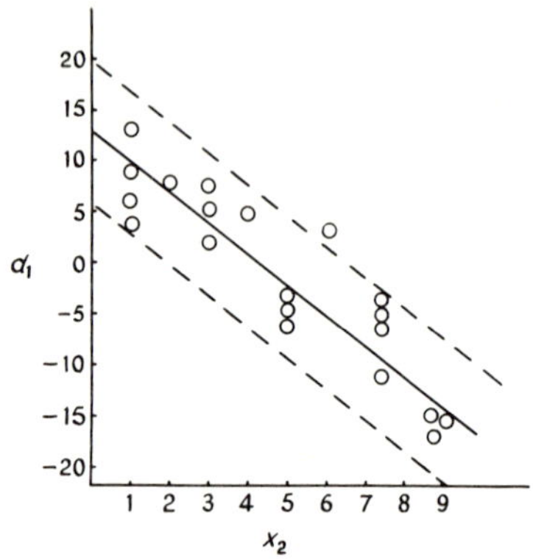

The limits shown can be used as approximate 95% limits for the final equation, i.e.

$$\text{Final } y = 6{\cdot}0 + 3{\cdot}45x_1 - 2{\cdot}67x_2 \pm 6$$

Note: the equation as calculated by the least squares technique is

$$y = 4{\cdot}30 + 4{\cdot}13x_1 - 2{\cdot}95x_2 \pm 4{\cdot}5$$

Calculation for each data point may be used to compare the predicted values. For example:

Point No.	*Actual*	*Predicted*
1	−6·6	−3·9
5	31·1	29·0
10	4·2	4·9
15	12·1	11·5
20	24·0	20·6

This method can be extended analogously to 3 or more x variables. Like the other 'quick and dirty' methods, it is an approximation to the best method, and should be used with judgement. Where time is at a premium and high-powered calculating equipment is not available, this approach may have some advantages, even though approximate.

Chapter VII discusses some of the possibilities available for correlations containing 3 or more variables, and where curvature exists in the relationships. These situations normally require computer handling.

PROBLEMS

1. Tests were run by two different laboratories on samples of emulsion polymer. Determine the least-squares line for estimating the expected result for laboratory II, given the value obtained by laboratory I, and the expected result for laboratory I given the value obtained by laboratory II. Are these two lines the same?

Laboratory I x	*Laboratory II* y
10	13
12	14
11	12
13	16
16	18
15	15
17	19
15	17
16	17
18	20

$$\Sigma x = 143 \qquad \Sigma y = 161$$

$$\Sigma x^2 = 2109 \qquad \Sigma y^2 = 2653$$

$$(\Sigma x)^2 = 20449 \qquad (\Sigma y)^2 = 25921$$

$$\Sigma xy = 2361$$

$$xy/n = 2302$$

$$\text{Difference} = 59$$

$$(\Sigma x)^2/n = 2045 \qquad (\Sigma y)^2/n = 2592$$

$$\bar{x} = 14{\cdot}3 \qquad \bar{y} = 16{\cdot}1$$

$$\Sigma x^2 - \frac{(\Sigma x)^2}{n} = 64 \qquad \Sigma y^2 - \frac{(\Sigma y)^2}{n} = 61$$

$$r = \frac{59}{\sqrt{(64)(61)}} = 0{\cdot}944$$

$$b_1 = 59/64 = 0{\cdot}922 \qquad b_2 = 59/61 = 0{\cdot}967$$

$$y = 16{\cdot}1 + 0{\cdot}922\,(x - 14{\cdot}3)$$

$$y = 2{\cdot}9 + 0{\cdot}922x$$

$$x = 14{\cdot}3 + 0{\cdot}967\,(y - 16{\cdot}1)$$

$$x = -1{\cdot}3 + 0{\cdot}967y$$

The second equation converts to $y = 1{\cdot}34 + 1{\cdot}03x$, which is not the same as the first.

INSTANT STATISTICS

'Birds of a feather . . .'

The current turmoil resulting from the recent findings on smoking versus cancer leads to logical as well as illogical discussion of the validity of the results. An inveterate smoker may say – 'I don't think the findings are conclusive – here are several cases where people who never smoked contracted lung cancer.' An air pollution crusader opines that people would not be sensitive to cigarette smoke were their lungs not constantly bombarded by other irritating substances. Or a 90-year-old gentleman who has smoked, chewed and taken snuff all his life may on X-ray show the lungs of a 20-year-old.

Do these exceptions prove the rule, or give the lie to the conclusions drawn. A critical examination of the data may help a trained observer reach his own decision; but those who have a vested interest may never admit the logicality of the conclusions.

Sometimes in a process study undertaken to determine the basic causes of some quality variation, the same turmoil may exist. We may show a relationship in a study of a month's data which seems to say that the colour stability of a product is related to the concentration of a metallic contaminant. The next month's data may fail to show this relationship, but may indicate an association with an organic contaminant. Another month's data may show both, or may fail to detect any effect due to either.

Over a period of time, more and more data are collected and, without computer techniques, the amount of time and effort in determining the over-all relationships may be prohibitive. In addition, the presence of these contaminants may have threshold effects, and linear regression may not apply. And there is always the one who says – 'Here's a good batch with high metal and high organic, that proves there's no relationship.'

A quick way of looking at a large mass of data to determine at least qualitatively the probability of some effect being associated with some variable is to divide the data into four segments as follows:

Variable low, effect low
Variable low, effect high
Variable high, effect low
Variable high, effect high

Then count the number of batches which fall into each category. On a plant problem the results might look something like below.

	Stability Good	Stability Poor	
Metal High	31	42	73
Metal Low	346	93	439
	377	135	512

	Stability Good	Stability Poor	
Organic High	27	55	82
Organic Low	350	80	430
	377	135	512

From this, the rough probabilities may be calculated.

Probability of a good batch with high metal	= 31/73	= 0·42
Probability of a good batch with low metal	= 346/439	= 0·79
Probability of a good batch with high organic	= 27/82	= 0·33
Probability of a good batch with low organic	= 350/430	= 0·81

Note: these probabilities are conditional, i.e., given that the metal is high, this is your chance of getting a good batch.

To test the effect of metal and organic together, proceed as follows:

	Metal High Organic High	Metal High Organic Low	Metal Low Organic High	Metal Low Organic Low
Stability Good	2	29	25	321
Stability Poor	12	30	43	50
	14	59	68	371
$\text{Probability}_{\text{good}}$	0·14	0·49	0·37	0·87

By this technique, the degree of association can be pictorially presented. The number of counts in each box will depend on the cutoff points between good and poor stability, and between the low and high levels of contaminants. These can be varied to investigate the effect of changing special limits and levels of control.

For the statistical purest, chi square tests may be used to determine significance levels, binomial distribution for confidence limits on the probabilities.

There is Always a Better Way

The quadrant sum test is recommended as a quick check on whether a correlation exists between two sets of measurements. It is a good method, although it relies entirely on the extreme values of the variables to detect a correlation. Every once in a while, one runs across a situation where the main bulk of the data appears to be correlated, but one or two of the extreme values fall out of line. In these situations, the quadrant sum test fails to detect the relationship.

A method which uses all the data is based on an exact test of a 2×2 contingency table and is recommended for cases like those mentioned above. It is also better than quadrant sum for large amounts of data (20 pairs or more).

The procedure is to plot the data in a scatter diagram as shown below, and to construct horizontal and vertical lines through the middle point (median) in each direction. This divides the data into four quadrants. Count the number of data points falling in any one quadrant (selecting the one containing the largest *or* smallest number of points). This total is compared to the table given below. If the smallest number is equal to or less than the corresponding number in the table, there is a correlation. Likewise, if the largest number is equal to or greater than the corresponding number in the table, there is a correlation (95% significance figures are given). *Note*: it is legitimate to try it both ways and use the approach that says the relationship is significant.

This method is contained in a good book by M.H. Quenouille entitled *Rapid Statistical Calculations* (Hafner, 1959).

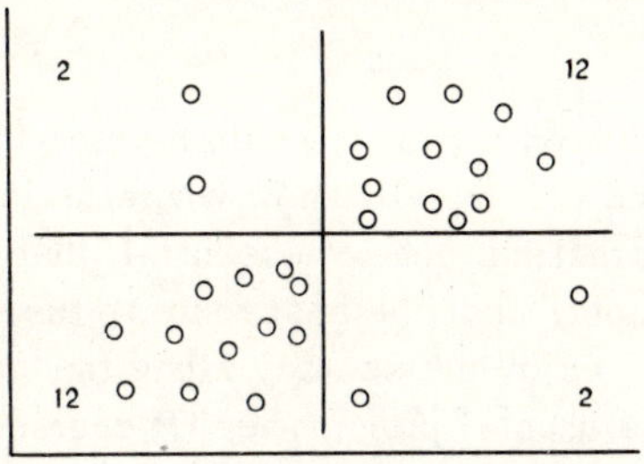

n = 28
Low Count = 2
High Count = 12

The quadrant sum test is spoilt by the extreme points in the upper-left and lower-right quadrants. Using the new method, reference to the table says that there is a significant relationship, using the low side or the high side of the table.

Table of significance levels

No. Points	*Low Side*	*High Side*
8 – 9	0	4
10 – 11	0	5
12 – 13	0	6
14 – 15	1	6
16 – 17	1	7
18 – 19	1	8
20 – 21	2	8
22 – 23	2	9
24 – 25	3	9
26 – 27	3	10
28 – 29	3	11
30	4	11
40	6	14
50	8	17
60	10	20
100	19	31

CHAPTER SIX

THE BEST-LAID PLANS . . .

When a person sets out on a trip, it would be nice for him to know where he is going and that he will know where he is when he gets there. It is also interesting, but not essential, that he knows how much it is going to cost, what the best route to the destination is (and he would like to enjoy the scenery along the way). The same is true about an experimental programme. Of course, it is conceivable that one can just start experimenting, and gradually find out what it is one would like to know, and to terminate the programme when the money runs out. But, I think we all agree that some planning is required.

There is a broad field of statistics which is devoted to the planning of experiments in a way that considers:

1. The purpose of the experimental programme.
2. The economics of achieving that purpose.
3. Ways of judging when the purpose is achieved.
4. Picking up a maximum of *new* knowledge along the way.

This is the science known as experimental design, and many books and papers, as well as word-of-mouth communication by those who know, are available to the interested research worker. Mathematical justification has been worked out for these designs, but in this test let us consider the patterns which seem to fit the particular problems the best. The basic patterns, which have been used most extensively in chemical research, either 'as is' or with appropriate modifications are:

1. *The Class of Factorial Designs* Square or rectangular patterns in two dimensions, cubic or rectangular parallelopipeds in three dimensions, and the analogous multidimensional figures in 4, 5 . . . to n dimensions.
2. *The Central Composite Designs* Similar to the factorial designs, but with lines radiating from the centre of the figure, perpendicular to the faces and terminating outside.
3. *The Simplex Designs* Straight lines in one dimension, triangles in two dimensions, tetrahedrons in three dimensions, etc.

In each of these patterns, the number of dimensions represents the number of *factors* or *variables* being studied. The value of the variable for a particular experimental sum is called the *level* of the factor. The measured quantity being used to evaluate the effect of these *variables* is known as the *response*.

All of these patterns can vary from very simple to very complex and, as the complexity increases, certain modifications can be introduced. In addition, the basic patterns can be combined as the need arises.

THE FACTORIAL DESIGNS

The simplest experimental design is one in which the effect of a single factor is investigated in two experiments. For example, the effect of method of agitation on the rate of solution of a solid in a liquid is being investigated. The experimental design is a one-factor factorial at two levels, three runs at each level, shown at the left.

This design may be used to determine which of two possible methods of operation might be carried forward as part of a more complicated study.

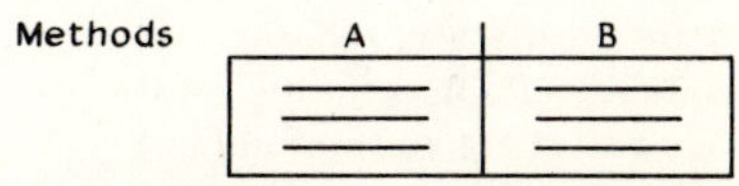

Factorial design – one factor at two levels. Each box represents one experimental condition. Enough runs should be run under each condition to assure that any observed difference is not just due to normal experimental variation.

The next design in degree of complexity is the two-factor, two-level factorial. This design is sometimes used to screen certain variables to determine whether their effect is important, and to estimate its direction, prior to larger scale quantification. It can be of advantage in determining what ranges of the variables should be studied.

This design permits calculation of the effect of each factor and also whether the effect of a factor is the same at each level of the

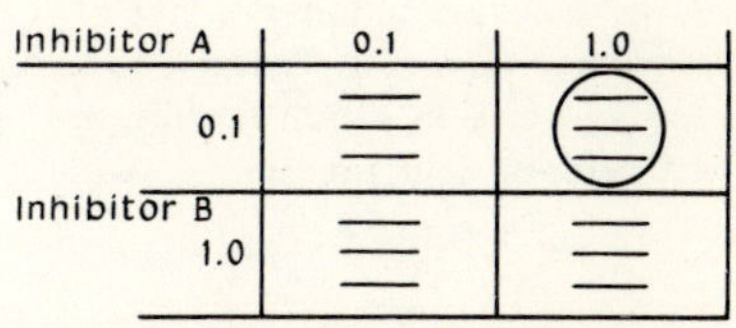

Factorial design – two factors at two levels. Each box represents one experimental condition. If we can be sure that the effect of concentration of each inhibitor is exactly the same at both concentrations of the other, only three out of four of the conditions needs to be run.

other factor. (This cross-effect is known as interaction.)

The two-level factorial design with three factors has $2^3 = 8$ experimental conditions. The four-factor design will have $2^4 = 16$ combinations. Beginning with the four-factor design through designs with more than four factors, the technique of fractional replication can be used. This consists in running a balanced fraction of the total number of possible combinations. A balanced fraction of a design is one in which the same number of experimental runs are conducted at each level of each factor.

		A_1		A_2	
		B_1	B_2	B_1	B_2
C_1	D_1	x	0	0	x
	D_2	0	x	x	0
C_2	D_1	0	x	x	0
	D_2	x	0	0	x

Factorial design – four factors at two levels. Running all 16 combinations is the full factorial. Running either the x's or the O's is called a 1/2 replicate.

As the number of factors increases, the economies of fractional replication become more evident. With six factors, a 1/4 replicate of the 64 possible combinations is not unusual and a 1/8 replicate of the 256 run eight-factor factorial is fairly common. Cochran and Cox [3] give a very extensive list of fractional factorial designs, including some in which factors are at three and four levels.

			A_1		A_2	
			B_1	B_2	B_1	B_2
C_1	D_1	E_1	x	0	0	x
		E_2	0	x	x	0
	D_2	E_1	0	x	x	0
		E_2	x	0	0	x
C_2	D_1	E_1	0	x	x	0
		E_2	x	0	0	x
	D_2	E_1	x	0	0	x
		E_2	0	x	x	0

Factorial design – five factors at two levels. Running all 32 combinations gives the full factorial. Either the x's or O's constitute a 1/2 replicate.

Factorial design – six factors at two levels. The boxes marked *X* represent 1/4 replicate of the complete $2^6 = 64$ factorial design.

			A_1				A_2			
			B_1		B_2		B_1		B_2	
			C_1	C_2	C_1	C_2	C_1	C_2	C_1	C_2
D_1	E_1	F_1	*X*						*X*	
		F_2				*X*		*X*		
	E_2	F_1		*X*						*X*
		F_2			*X*		*X*			
D_2	E_1	F_1				*X*		*X*		
		F_2	*X*						*X*	
	E_2	F_1			*X*		*X*			
		F_2		*X*						*X*

Factorial designs with more than two levels of the factors are quite common, and mixed factorial designs in which the several factors have different numbers of levels might fit certain experimental requirements. The fractional replication of designs of this type is somewhat hazardous, since balanced arrangements are hard to come by. Fractional replicates of experimental designs in which all factors* are at the same number of levels can be partially replicated in fractions whose denominators are multiples of the number of levels. These designs are the so-called *Latin square designs*.

	A_1			A_2		
	B_1	B_2	B_3	B_1	B_2	B_3
C_1						
C_2						
C_3						
C_4						
C_5						

Factorial design – one factor at two levels, one factor at three levels and one factor at five levels. Number of runs $= 2^1 \times 3^1 \times 5^1 = 30$.

The National Bureau of Standards has published two booklets: *Fractional Factorial Designs for Experiments with Factors at Three Levels* and *Fractional Factorial Designs for Experiments with Factors at Two and Three Levels*. Some of the possible experiments covered include:

1/2 replicate of 5 factors at 2 levels, 1 at 3 levels

1/4 replicate of 4 factors at 2 levels, 3 at 3 levels

	A_1	A_2	A_3
B_1	C_1	C_2	C_3
B_2	C_2	C_3	C_1
B_3	C_3	C_1	C_2

Latin square design – three factors at three levels. Run the indicated experiment at the levels of A, B, and C shown in the box. This is a 1/3 replicate of a $3^3 = 27$ factorial design.

* 3 or more

1/6 replicate of 2 factors at 2 levels, 6 at 3 levels

1/18 replicate of 2 factors at 2 levels, 7 at 3 levels

and even

3/4 replicate of 4 factors at 2 levels, 1 at 3 levels

	A_1	A_2	A_3	A_4
B_1	C_1	C_2	C_3	C_4
B_2	C_2	C_3	C_4	C_1
B_3	C_3	C_4	C_1	C_2
B_4	C_4	C_1	C_2	C_3

Latin square design – three factors at four levels. This is a 1/4 replicate of a $4^3 = 64$ factorial design.

Cochran and Cox [3] have examples of Latin squares (three factors) up to twelve levels; and Graeco Latin squares (four factors) up to twelve levels.

The fractional factorial designs, including the Latin squares, are generally used for screening possible experimental variables in order to find which are the most important for further study. Their use is subject to some fairly severe assumptions which should be known and taken into consideration when interpreting the data.

THE CENTRAL COMPOSITE DESIGNS

The factorial designs are not limited to experiments to determine the existence and relative importance of the experimental variables. The information gained from these designs can often be used to provide equations for quantifying the efforts of these variables on some response. When an experimental programme is undertaken with the calculation of an equation in mind, another set of experimental designs has been found to be more efficient for this purpose. These are the so-called central composite designs. The central composite designs consist of the basic factorial design (either complete or fractionally replicated), with some additional experiments in a symmetrical pattern. In the two- and three-factor designs, these additional points can be visualized as forming a star, with a point at the centre of the experimental space, and extreme points, outside the factorial design, for each factor at the midpoint of all the others.

The two-factor central composite as shown consists of the 2^2 factorial designated as

$$\left.\begin{aligned} x_1 &= \pm 1 \\ x_2 &= \pm 1 \end{aligned}\right\} \text{all four combinations.}$$

In addition, there is a centre point at $x_1 = 0$, $x_2 = 0$ and star points at $x_1 = 0$, $x_2 = \pm\sqrt{2}$, $x_2 = 0$, $x_1 = \pm\sqrt{2}$. In this design, it is suggested that the centre point be run five times to provide an estimate of experimental error and to determine how well the calculated equation fits the data.

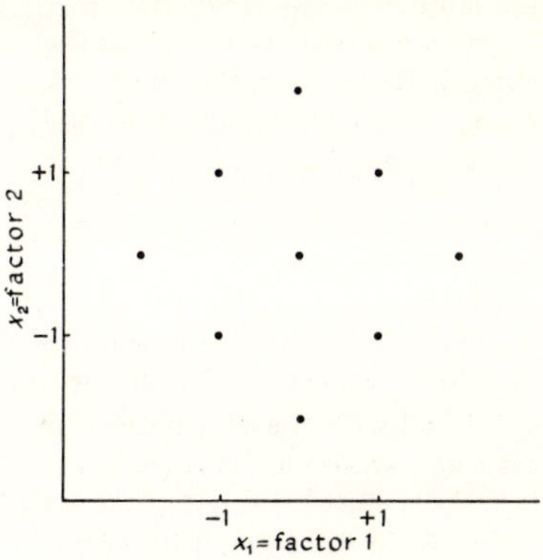

Two-factor central composite.

The three-factor central composite may be represented as a cube, with a centre point, experiments at each vertex and at the ends of axes radiating out from the centre through the middle of each face. The vertices will be designated as +1 or −1 for each factor. It is suggested that six runs be made at the centre. This design can be represented as indicated below the diagram.

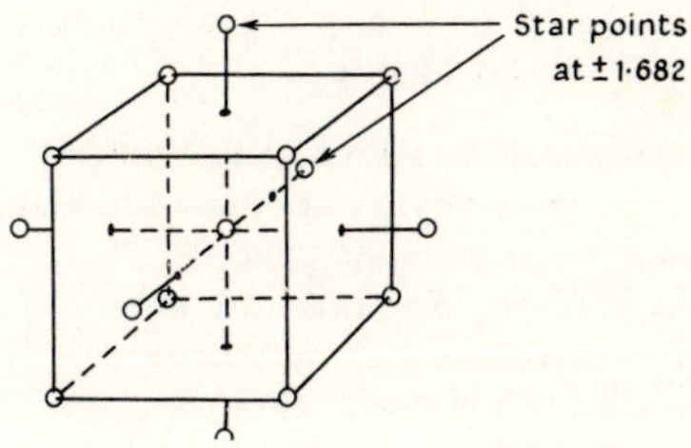

Three-factor central composite.

Run No.	x_1	x_2	x_3
1	−1	−1	−1
2	−1	−1	+1
3	−1	+1	−1
4	−1	+1	+1
5	+1	−1	−1
6	+1	−1	+1
7	+1	+1	−1
8	+1	+1	+1
9	0	0	0
10	−1·68	0	0
11	+1·68	0	0
12	0	−1·68	0
13	0	+1·68	0
14	0	0	−1·68
15	0	0	+1·68

This notation is commonly used in the literature.

In setting up the design, we decide the range over which we would like to study the variables and then set the levels for each experimental condition to conform with the design. For example, if x_1 = temperature, x_2 = pressure, and x_3 = catalyst concentration, we might assign the following experimental levels.

	−1·68	−1	0	+1	+1·68
Temp.	83·2°	90°	100°	110°	116·8°
Press.	31·2	35	40	45	48·4
Cat.	0·033	0·040	0·050	0·060	0·067

Although the star point patterns indicated above are optimal from the mathematical viewpoint, a common practice is to set up the outriggers at ±2 rather than ±1·41 or ±1·68 for the two- and three-dimensional cases.

A five-factor central composite design consists of the five-factor, two-level factorial, with the centre point and with the star pattern in all five variables. This would ordinarily call for running 32 + 10 + + 1 = 43 conditions, with some replication at the centre. Usually, the half replicate of the factorial plus the star points and centre is enough to give an adequate picture of the relationships.

Many modifications of the central composite design are possible and may be used to fit specific situations. For example, one of the variables may only have two possible levels. In this case, the portion of the star pattern for this variable would be omitted and the star pattern for the other variables might be run at a selected level of the variable in question. In other cases, the outrigger points may be meaningless for a particular variable, and only the factorial and centre points used.

The rotatable feature of the central composite designs makes it possible to complete a balanced portion of the design, evaluate the results and possibly shift the design to another area in some of the variables. The shift in area of interest pivots on some of the runs already obtained and these become part of the new experimental design.

Five-factor central composite, half replicate and star points.

Run No.	x_1	x_2	x_3	x_4	x_5
1	−1	−1	−1	−1	+1
2	−1	−1	−1	+1	−1
3	−1	−1	+1	−1	−1
4	−1	−1	+1	+1	+1
5	−1	+1	−1	−1	−1
6	−1	+1	−1	+1	+1
7	−1	+1	+1	−1	+1
8	−1	+1	+1	+1	−1
9	+1	−1	−1	−1	−1
10	+1	−1	−1	+1	+1
11	+1	−1	+1	−1	+1
12	+1	−1	+1	+1	−1
13	+1	+1	−1	−1	+1
14	+1	+1	−1	+1	−1
15	+1	+1	+1	−1	−1
16	+1	+1	+1	+1	+1
17	−2	0	0	0	0
18	+2	0	0	0	0
19	0	−2	0	0	0
20	0	+2	0	0	0
21	0	0	−2	0	0
22	0	0	+2	0	0
23	0	0	0	−2	0
24	0	0	0	+2	0
25	0	0	0	0	−2
26	0	0	0	0	+2
27	0	0	0	0	0

The central composite designs are very handy if it is necessary to run an experiment in several different blocks. There may be many reasons for doing this. For example, we may have only enough raw material to run a limited number of experiments. Or a test method may be such that we can run a given number of samples at a time. Or we may wish, after a quick look at the results, to change to some other range of variables.

Blocking is done by running a set of experiments in balanced blocks and correcting the results for the differences (if any) between blocks prior to or in the process of analysing the results. In a four-factor central composite design, the experiment may be split into three equal blocks as indicated.

Block I Half replicate of factorial design and centre point.

Block I

Run No.	x_1	x_2	x_3	x_4
1	−1	−1	−1	−1
2	−1	−1	+1	+1
3	−1	+1	−1	+1
4	−1	+1	+1	−1
5	+1	−1	−1	+1
6	+1	+1	+1	−1
7	+1	+1	−1	−1
8	+1	+1	+1	+1
9	0	0	0	0

Block II Other half replicate of factorial design and centre point.

Block II

1	−1	−1	−1	+1
2	−1	−1	+1	−1
3	−1	+1	−1	−1
4	−1	+1	+1	+1
5	+1	−1	−1	−1
6	+1	−1	+1	+1
7	+1	+1	−1	+1
8	+1	+1	+1	−1
9	0	0	0	0

Block III Star points and centre point.

Block III

1	−2	0	0	0
2	+2	0	0	0
3	0	−2	0	0
4	0	+2	0	0
5	0	0	−2	0
6	0	0	+2	0
7	0	0	0	−2
8	0	0	0	+2
9	0	0	0	0

THE SIMPLEX DESIGNS

Some experimental problems involve mixtures of ingredients and the independent variables are the proportions of the various materials present in the mixtures. In general, the factorial and central composite designs are not applicable except in certain specific cases. The reason is that the design must be set up in such a way that the sum of the proportions of the ingredients present must always be equal to one (or to some constant percentage of a larger mixture).

The designs for this kind of experimentation can be built up similarly to the factorial designs, starting with the simple case in which only two materials are mixed together. The effective number of variables will be one less than the number of ingredients.

Two-component design

Experiment No.	*Component A*	*Component B*
1	1·00	0·00
2	0·75	0·25
3	0·50	0·50
4	0·25	0·75
5	0·00	1·00

The two-component design has effectively only one variable. Once the proportion of component A is set, the proportion of B is fixed by the relationship. Proportion (A) + proportion (B) = 1·00

Three-component design

Experiment No.	*Component A*	*Component B*	*Component C*
1	1·00	0·00	0·00
2	0·00	1·00	0·00
3	0·00	0·00	1·00
4	0·33	0·33	0·33
5	0·50	0·50	0·00
6	0·50	0·00	0·50
7	0·00	0·50	0·50

(See illustration below.)

The three-component design has effectively two independent variables, but the flexibility is increased by the fact that the levels of any two can be established independently of each other.

The number of experiments to be run in a multicomponent mixture problem will depend on several factors. These include:

1. How many components are to be looked at?
2. Are the mixture properties linear with component concentrations?
3. Do we expect a synergistic effect?
4. Is there any particular group of mixtures of special interest?

5. Are there restrictions on the concentrations of any of the components?

Three-component mixtures can be represented on triangular coordinates and the reasoning used in three-component designs can be extended to four or more components. Some typical three-component designs follow.

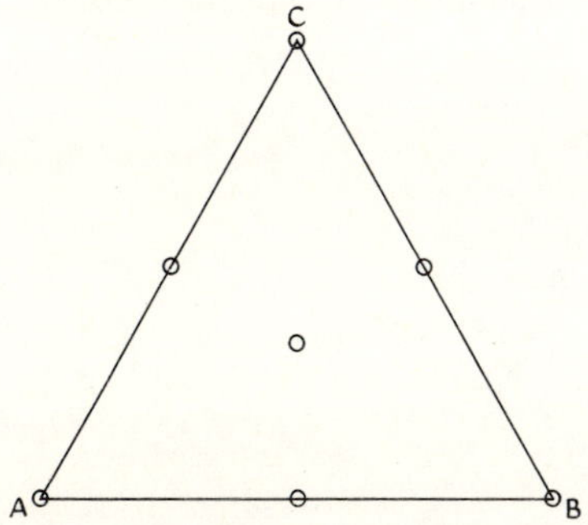

The apex marked A represents 100% of component A; B represents 100% of component B, etc. A point on the side midway between B and C represents a 50–50 mixture of components B and C. The circles represent the three-component design described above. This design is generally used when the area of interest covers all possible combinations of A, B, and C, and the properties bear fairly simple relations to the percentage of each component present.

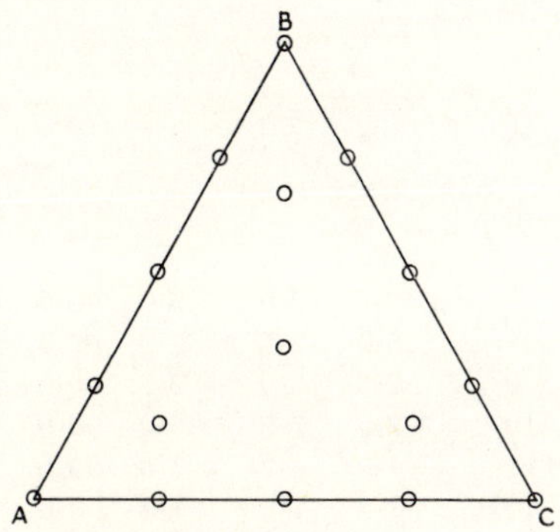

This design can be useful when the area of interest represents all possible combinations, but the relationship is expected to become somewhat complex or some synergism is expected. In addition to the 1:0:0, 1/2 : 1/2 : 0 and 1/3 : 1/3 : 1/3 combinations, all 1/4 : 3/4 : 0 and 1/6 : 1/6 : 2/3 combinations are included.

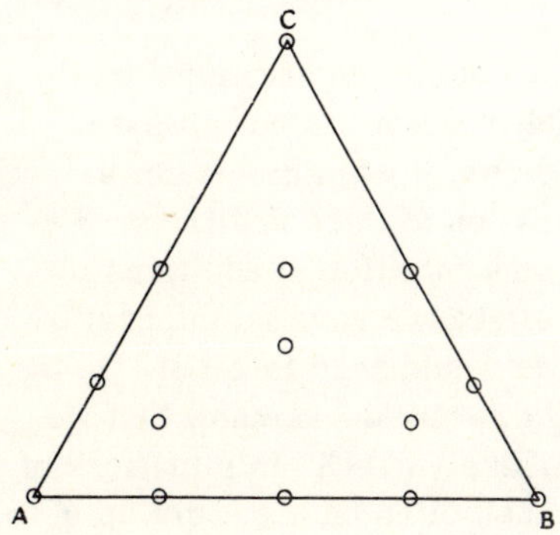

Where it may not be feasible or economical to include more than 50% of ingredient C, and where the response is still expected to be complex, this truncated design may be used. Here the points containing more than 50% C have been omitted and a 25 : 25 : 50 ratio of A : B : C has been added.

In some multicomponent mixture problems, it may be desirable to hold one component at some constant level and vary the others. In this case, we ignore the constant component and use the Simplex designs in the other components, assuming the constant total of these as 100% of the mixture. In the case of an oil additive, for example, we may wish to specify a mixture of additives which add up to 10% of the total composition. A possible design for a study of this type is given below:

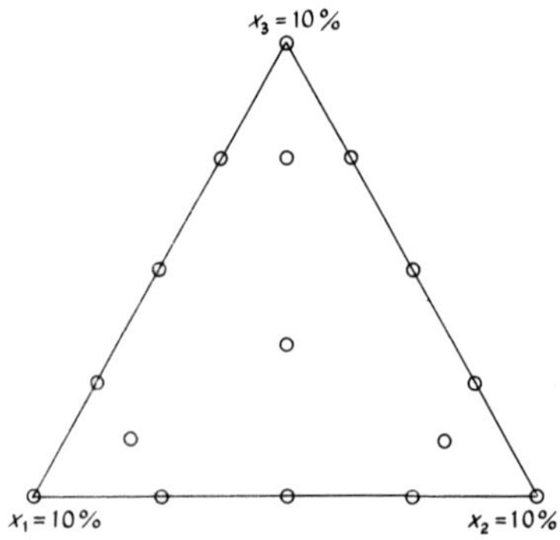

Four-component lube mixture – base stock = 90%

Run No.	x_1	x_2	x_3	x_4	*Run No.*	x_1	x_2	x_3	x_4
1	10·0	0·0	0·0	90·0	7	0·0	0·0	5·0	90·0
2	0·0	10·0	0·0	90·0	8	5·0	2·5	2·5	90·0
3	0·0	0·0	10·0	90·0	9	2·5	5·0	2·5	90·0
4	3·3	3·3	3·4	90·0	10	2·5	2·5	5·0	90·0
5	5·0	5·0	0·0	90·0	11	1·2	1·2	7·6	90·0
6	5·0	0·0	5·0	90·0	12	1·2	7·6	1·2	90·0
					13	7·6	1·2	1·2	90·0

One of the limitations to the Simplex design in additive experimentation is that the desired total additive level is not always known. For example, we may be interested in obtaining a certain effect on a base stock using a combination of three additives. For cost reasons, we may limit the total concentration of additives to 6%. If we do this and fail to get the effect, we have no information to estimate how much more additive we would need to get it. On the other hand, if we do get the effect at 6%, how do we know that we would not have obtained it less expensively at 4%. In situations of this kind, a central composite design can often be used, set up in such a way that the extremes of the design lie within certain limits. Or several complementary Simplex designs, each at a different level of concentration may be employed.

If the additive concentration is very low, as is the case sometimes with antioxidants, detergents, and lubricity additives, the central composite design may again be preferable to the Simplex.

EXPERIMENTAL PLANNING IN GENERAL

The foregoing experimental designs are merely examples of the many tried and true patterns available to the experimenter. Of course, the best approach would be to design each experiment to fit the situation at hand and, in most cases, this can be done by judiciously selecting one of the standard patterns.

To fit the proper experimental design to the situation, the following outline and check list may be valuable:

1. Statement of objectives.
   (a) Ultimate goal.
   (b) Specific information needed.
   (c) Proposed use of the information.
2. Description of experiment.
   (a) Variables to be studied.
   (b) Ranges of variables.
   (c) Number of runs.
   (d) Materials and equipment to be used.
   (e) Recognition and control of external variables.
   (f) Methods of measurement to be used.
3. Analysis and interpretation of results.
   (a) Qualitative screening.
   (b) Selection of most important factors.
   (c) Quantification of effects.
   (d) Optimization.
4. Feedback and reassessment.
   (a) Further objectives.
   (b) Additional experiments needed.

Check list

1. Are we looking for fundamental information or are we trying to

'spot the winner' as soon as possible?

2. What are the variables?
3. Which will be held constant?
4. Which will be deliberately changed?
5. Are there any variables which we will permit to change at random, but will measure?
6. Are all the desired conditions capable of attainment?
7. Do we need to run a control or standard?
8. Are we trying to quantify the effects of the variables or merely determining their relative importance?
9. How small an effect are we interested in detecting?
10. How many experiments do we need to run?

The following table will help determine the number of experimental runs required to detect an effect of some given magnitude. If we have some feel for the reliability of our results and can express this as a standard deviation, we can also express the desired effect in standard deviation units. This is a realistic approach to establishing the size of the experimental programme and should be geared to the effect which we think will be the smallest, provided it is still of practical importance. If an effect below a certain magnitude is of no practical importance, then we should not worry whether the experiment is capable of detecting it.

Number of experiments required to detect a given difference at 95% confidence

	Probability of detection		
Difference	0·80	0·90	0·95
5σ	2	2	3
4σ	2	2	4
3σ	3	3	5
2σ	4	5	9
$1{\cdot}5\sigma$	7	9	15
$1{\cdot}0\sigma$	13	18	33
$0{\cdot}8\sigma$	19	26	47
$0{\cdot}6\sigma$	33	45	>50
$0{\cdot}5\sigma$	50	>50	>50

σ is a measure of total experimental error.

Some typical examples of experimental designs and the situations to which they apply follow.

1. *Determination of test error*

Two samples are taken from each of two lots of the material normally tested. Each sample is tested in duplicate by each of two analysts on three separate days. If a preparation of the sample is required prior to testing, such a preparation should be made by each analyst on each sample each day, and two determinations run on the same preparation.

More or fewer levels of each factor or more tests on each sample may be run as required.

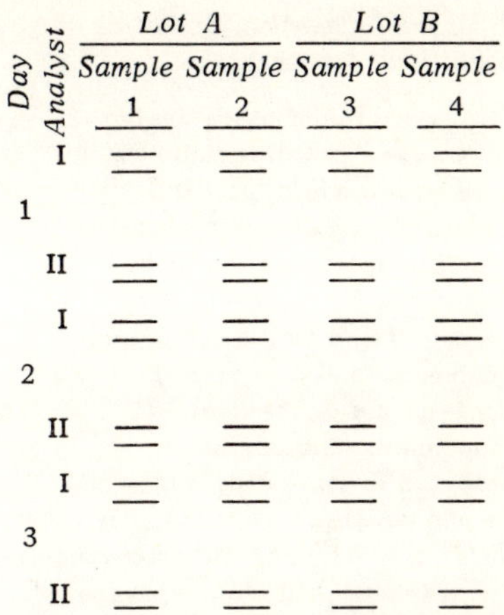

For established tests run under prescribed standard conditions.

2. *Effect of conditions on level and variability of results*

This approach is used in deciding what operating conditions should be used in a new test. Three determinations are run at each set of conditions, and we are interested in selecting conditions which have the smallest variation between tests run at that condition and which also show the greatest differences between materials. Economy can often be affected by making this study a part of a larger study in which other factors are also investigated.

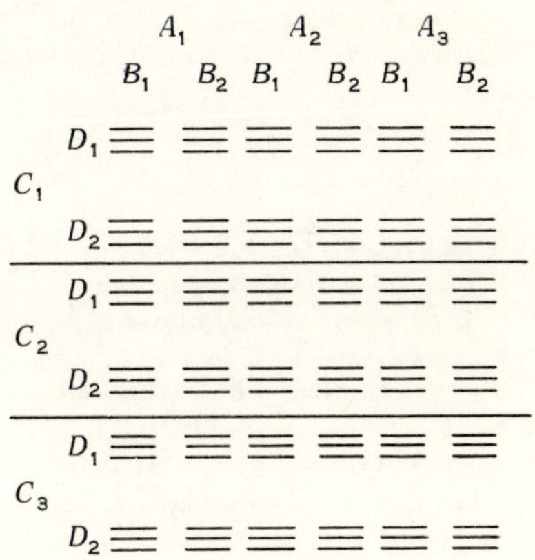

Example:

A = molding time

B = molding temperature

C = milling time

D = milling temperature

3. *Establishment of operating conditions for a process*

The central composite design shown here is tailor made for determining *yield, quality, throughput, cost,* etc., for all feasible operating conditions. Equations for each response can be calculated and operating conditions selected to give the best combination of the desired responses. The condition selected for operating may well be one which was not actually run in the experimental programme; for example, in the process shown we may decide to operate at 60 min, 100°, and a catalyst concentration of 0•10.

	Time	*Temperature*	*Catalyst Concentration*
1	70	90	•05
2	70	90	•10
3	70	110	•05
4	70	110	•10
5	100	90	•05
6	100	90	•10
7	100	110	•05
8	100	110	•10
9	85	100	•075
10	50	100	•075
11	120	100	•075
12	85	80	•075
13	85	120	•075
14	85	100	•025
15	85	100	•125

This design is also used in experiments seeking a maximum or minimum.

4. *Existence of effects*

In many systems, there may be just a few variables which have a large effect on the property being measured. If we can make this assumption, we may be able to select these variables from a number of others by a relatively small number of experiments. A design of the type shown can be used in which two levels, a high and low level, of each factor can be selected. In the design, 0 represents the low level and 1 the high level of each factor.

	A	*B*	*C*	*D*	*E*	*F*	*G*
1	0	0	0	1	1	1	0
2	0	0	1	1	0	0	1
3	0	1	0	0	0	1	1
4	0	1	1	0	1	0	0
5	1	0	0	0	1	0	1
6	1	0	1	0	0	1	0
7	1	1	0	1	0	0	0
8	1	1	1	1	1	1	1

Y =	electrical property of product	
A =	time	60, 90 min
B =	temperature	90°, 110°
C =	catalyst	0•05, 0•10
D =	acid treatment	no, yes
E =	additive A	0, 1
F =	additive B	0, 1
G =	additive C	0, 1

Running these eight should pick out any really strong factors that exist. *However, they should be confirmed by subsequent runs.*

5. *Optimum composition of a mixture*

This example shows a design set up to determine the thermal stability of a mixture containing 3·85% of a total of three additives plus a constant amount (1·15%) of a fourth. The design is set up as a Simplex design in the three variable additives. *Note:* in no case was any ingredient present alone; the main area of interest in ingredient C was 0·35 to 1·25%.

Run no.	*Additive A(%)*	*Additive B(%)*	*Additive C(%)*
1	0·00	3·50	0·35
2	0·00	2·60	1·25
3	1·15	2·35	0·35
4	0·85	1·75	1·25
5	2·35	1·15	0·35
6	1·75	0·85	1·25
7	3·50	0·00	0·35
8	2·60	0·00	1·25
9	1·50	1·50	0·85
10	1·00	1·00	1·85
11	1·92	1·93	0·00

6. *Screening of ingredients in pairs*

The materials to be tested are divided into two groups, one called the *primary group* and one called the *secondary group*. Each experiment will consist of one primary material plus a secondary material at some concentration. In the example shown, five primary, five secondary, and five concentrations were screened in twenty-five runs. Five additional secondary materials with the same five primaries were then screened in a similar design.

Secondary \ Primary	A	B	C	D	E
1	2%	8%	4%	0%	6%
2	8%	4%	0%	6%	2%
3	4%	0%	6%	2%	8%
4	0%	6%	2%	8%	4%
5	6%	2%	8%	4%	0%

5 × 5 Latin square

Numbers in blocks indicate the concentration of the secondary material.

In looking at the complete experimental programme, it is often useful to think in terms of combinations of experimental designs. For example, many variables might be looked at by using a screening design or a fractional factorial, then the results confirmed and interactions detected by using a complete factorial. Then, finally, setting most of the variables at some predetermined level, we may optimize over the others by using the central composite design.

Synthesis-application programmes may be carried out by using a simple design in some basic variables to decide what compounds to synthesize. Then each of these can be checked for optimum performance under varying application conditions.

Plant experimentation will require a fairly large-scale plan which is partitioned into balanced blocks to fit in with production needs, ability to change conditions, etc. Cochran and Cox [4] give balanced designs which are broken up into small, easy to handle, blocks.

EVOLUTIONARY OPERATION

A handbook of practical statistics would not be complete without at least a mention of evolutionary operation (EVOP). Although strictly speaking evolutionary operation is a method for operating plant processes, and perhaps less applicable to chemical research, a research worker may never know when he will be called upon to help the production department out of a tight spot.

Evolutionary operation is a system for obtaining information on a going process, which will permit stepwise improvement of the process without risking a serious deviation from standard operating conditions. The method consists in observing a property or properties of the material such as quality, yield, etc., as a function of two or more operating variables.

In research experimentation, we normally try to improve the sensitivity of the experiment by reducing the experimental error and by taking the data over a wide range of the independent variables. In EVOP, the experimental error is accepted 'as is', the range of the variables is kept quite narrow and the experiments are run sequentially until enough data are obtained to permit a decision. The computations are kept simple at each step, and the information gained at each step is used along with the cumulative information from previous runs.

A detailed discussion of the procedure can be found in two papers [10] and [11]. Any research worker who has occasion to monitor the operation of a production process will wish to become familiar with the advantages and shortcomings of EVOP.

INSTANT STATISTICS

From here to infinity

One may be confronted with a problem which has an infinite number of possible solutions. In the business of blending – gasolines, fertilizers, animal feed stocks, functional fluids, oil additives, plasticizers – this kind of problem often occurs. Twenty different materials may be available, having differing properties and differing costs. Some of the materials will meet the required specifications in certain properties, some in others. Usually we find that the really low-cost material, which we would like to see included, does not meet any of the specifications. It is difficult to know where to begin. Linear programming may be able to help.

Linear programming is a mathematical procedure which permits you to optimize some quantity, while holding others within certain limits. For example, we may wish to minimize the cost of a multi-component formulation, while insisting that the combination of the several materials has certain required properties.

In an outlandishly oversimplified situation, we may wish to obtain a minimum cost oil additive which meets the following conditions:

Metal content	≥ 30%
Soap content	≥ 30%
Compatibility with oil	≥ 20%

We have four additives available, with the following properties:

Additive	*Metal* (%)	*Soap* (%)	*Compatibility* (%)	*Cost*
A	42	27	24	\$1·50
B	65	15	20	\$1·25
C	25	42	16	\$1·00
D	40	40	20	\$2·00

(Note that additive D meets the required specifications alone, but at a cost of \$2·00/lb.)

The linear programming technique proceeds to evaluate the cost of certain formula combinations at critical points, for example the maximum amounts of the cheaper materials permitted by the specifications; the minimum amounts of the more expensive materials required to meet the specifications, etc. In this problem, the answer obtained is:

Material A = 38·5%
Material B = 23·0%
Material C = 38·5%

which contains 40·76% metal, 30% soap and has a compatibility in oil of 20%. The cost of this formulation is \$1·25/lb.

There are certain limitations to the use of this technique, but if you have this kind of problem, the possibilities are worth exploring.

This is especially true if there are many available materials and many specifications.

Bayesian or 'crystal ball' statistics

Students of the occult generally agree that much of fortune-telling has nothing to do with the ability of the practitioner to forecast the future. On the contrary, the uncanny ability of the gypsy woman to tell the customer what he or she would like to hear is based on a shrewd ability to assess the facts contributed by the customer and to construct these facts into a plausible sequence of possibilities. The dark tent, crystal ball, and fancy costuming are merely trimmings to enhance the effect.

Of late, a rather vociferous group of statistical practitioners have been advocating what they call the Bayesian approach to decision-making – both in the area of management decisions and in assessing the meaning of experimental information. Some have even gone so far as to suggest 'experimentation without data', a systematic approach to quantifying existing knowledge as a foundation for more decisive experimentation.

The term Bayesian comes from Thomas Bayes, an 18th-century Presbyterian minister, who wrote *An Essay Towards Solving a Problem in the Doctrine of Chances*. The meat of the essay was that given some prior knowledge of the state of nature, the probability with which a decision could be made in a particular set of circumstances would be different than if the prior knowledge were not available.

Let us suppose there are three identical boxes, each containing two balls. Suppose also that it is known that one box contains two black balls, one contains a black and a white ball, and one contains two white balls. Now, a box is sampled, without looking, and a black ball is obtained. What is the probability that it came from the box containing two black balls?

Before sampling, the probability that any given box contains two black balls is 1/3. This is the existence probability. We also know that the probability of drawing a black ball from each of the three boxes is 1, 1/2, and 0, respectively. These are the productive probabilities. The product of the productive probability and the existence probability for each box is divided by the sum of all such products, and the results are the posterior probabilities for the hypothesis. The following table shows the calculations.

Hypothesis box is	*Existence probability*	*Productive probability*	*Product*	*Probability that hypothesis is correct*
BB	1/3	1	1/3	2/3
BW	1/3	1/2	1/6	1/3
WW	1/3	0	0	0

In other words, if we decide that the box contains two black balls, after drawing the sample, we will be right 2/3 of the time.

Perhaps there is a rather tenuous connection between this simple example and 'experimentation without data'. The basic concept is that the presence of the existing knowledge of the state of nature (in this case the distribution of the black and white balls) helps to make a better decision than if we were forced to assume, for lack of information, that each box contained one white and one black ball.

In a manufacturing operation, knowledge of the distribution of quality in a particular process makes it possible for us, on a given batch, to decide:

1. To ship without sampling.
2. To reject without sampling.
3. To sample and calculate posterior probabilities.

The cost of each line of action depends on the cost of sampling, reworking and of shipping unsatisfactory material. If the prior distribution were not known, decision would have to be based on the sample and a minimum cost decision procedure might not be reached.

In experimentation, prior knowledge can be quantified by making use of experience, theory, intuition, and all the scientific and pseudoscientific means for estimating what would happen in a given situation. This can be quantified by computer techniques into a prior probability distribution. Knowing this distribution, experimentation can proceed sequentially and selectively, minimizing total cost and maximizing the contribution of past experience. The crystal ball is not necessary, but neither is excessively extensive experimentation.

CHAPTER SEVEN

WHAT DOES IT ALL MEAN?

The experimental design is the key to getting the most out of the experimental programme. As Mother used to say, 'Well begun is half done.' Collecting the data according to a balanced, logical, prearranged pattern in many cases will make the interpretation of the results obvious, without resorting to fancy mathematical treatment. However, we are always aware of the presence of error: variation in operating conditions, errors of measurement, existence of extraneous random variation. The question arises when we look at the data in various logical arrangements – how much chance is there that we are kidding ourselves when we place this or that interpretation on the results?

The statistical methods available make use of the pattern and magnitude of the differences among our experimental results, to tell us what is the chance of being wrong in drawing certain conclusions. There are many techniques available, but by far the majority of applications in chemical experimentation may best be treated by *analysis of variance* and *regression analysis.*

ANALYSIS OF VARIANCE

The analysis of variance lends itself best to balanced factorial designs, whether complete, partially replicated, or otherwise modified. The concept of balance simplifies the calculations tremendously. There are ways of coping with missing data, unequal replication under various conditions, and even some lack of orthogonality in the design, but these methods seem to involve more calculation than the data may deserve. The analysis of variance is a procedure which makes it possible to compare the effects of the variables being studied, first independently of the effects of all other variables, and second in all possible combinations with one another. Sometimes the effect of a variable within a given level of another variable is calculated. Some of the questions answered by the analysis of variance are:

1. Does this factor have an effect? (That is, does a change in the level of this factor result in a corresponding change in the response? cf. p. 34).
2. How sure can I be that this apparent effect is real?
3. Does the effect of a factor depend upon the level (or presence) of another factor?
4. How sure can I be that this is the case?
5. Is the effect linear over the range of the variable studied?
6. How sure can I be that the effect is linear (quadratic, cubic, quartic or a combination of these)?
7. How much of the total variation observed is due to each factor studied?
8. After all factors and interactions have been accounted for, how much variation is left unexplained?
9. What are the confidence limits of each of the observed effects?
10. Does the effect exist over the entire range of the variable studied, or only over a portion of the range?

And many other similar questions.

The techniques involved in the analysis of variance will differ somewhat for each particular experimental design. Brownlee (*Industrial Experimentation*) [1] handles many of these variations in a readable 'cookbook' style and is recommended for the practitioner who wants to know 'how to', but not necessarily 'how come?'. More of the background theory can be found in Davies [4], Dixon and Massey [5], and Cochran and Cox [3]. For our purposes, let us go through a typical example.

Procedure

Consider an experiment run to investigate the effects of time, temperature, and catalyst on the percentage of residual monomer in a simple polymerization reaction. We may be looking for the best way in which to quantify these effects, and would like to get the feel of the operation and some idea as to how well we can duplicate our results. A $2 \times 2 \times 3$ factorial is set up,

Calculations

	1 hour		2 hours		3 hours	
Catalyst	0·5	1·0	0·5	1·0	0·5	1·0
100°	5·4	2·3	3·8	2·0	2·1	0·8
	4·8	1·9	3·6	1·6	1·9	0·6
120°	4·2	2·2	3·5	1·0	1·6	0·4
	4·2	1·4	3·1	1·0	1·4	0·0

The results of an experimental programme. The numbers represent the percentage of monomer in two independent runs under identical conditions.

with duplicate runs at each combination of conditions. The results are as shown.

Form the *two- and three-way tables* as illustrated. These tables serve a double purpose. The totals are used in the calculations, and the tables themselves show a picture of the effects which is equivalent to graphical plotting.

Table 1 is formed by adding the duplicate values in each block

Two- and three-way tables

Table 1

Time × temperature × catalyst

	1 hour		2 hours		3 hours	
Catalyst	0·5	1·0	0·5	1·0	0·5	1·0
100°	10·2	4·2	7·4	3·6	4·0	1·4
120°	8·4	3·6	6·6	2·0	3·0	0·4

Table 2 is formed by adding all the values corresponding to a given combination of time and temperature. Totals for each time, each temperature, and for all observations are crossfooted as shown.

Table 2

Time × temperature

	1 hour	2 hours	3 hours	Total
100°	14·4	11·0	5·4	30·8
120°	12·0	8·6	3·4	24·0
Total	26·4	19·6	8·8	54·8

Table 3 is formed by adding all the values corresponding to a given time and catalyst concentration. Crossfooted values can be checked for accuracy with those in Table 2.

Table 3

Time × catalyst

	1 hour	2 hours	3 hours	Total
0·5	18·6	14·0	7·0	39·6
1·0	7·8	5·6	1·8	15·2
Total	26·4	19·6	8·8	54·8

Table 4 is formed by adding all the values corresponding to a given temperature and catalyst concentration. Crosschecking of all variables is now possible. *Note*: the order of magnitude of the different effects can be seen by inspection of the tables. Also any

Table 4

Temperature × catalyst

	100°	200°	
0·5	21·6	18·0	39·6
1·0	9·2	6·0	15·2
Total	30·8	24·0	54·8

interdependence of the effects will also be apparent.

The value of forming these tables cannot be overemphasized. Sometimes the results are sufficiently obvious to preclude the need for full calculation of the analysis of variance.

Now we are ready for the mathematics.

Procedure	*Calculations*
1. Obtain the sum of all the original observations. Call this GT or grand total.	1. GT = 54·8
2. Obtain the sum of the squares of all the original observations. Call this SSI.	2. SSI = 174·70
3. Determine $(\mathrm{GT})^2/n$ where n is number of original observations. This is sometimes called the correction factor CF.	3. $\mathrm{CF} = (\mathrm{GT})^2/n = 125{\cdot}13$
4. Subtract the CF from the SSI. The result is called the total sum of squares or TOTAL SS.	4. TOTAL SS = SSI − CF = 174·70 − 125·13 = 49·57
5. Square the total (t) of the observations at each time, divide by the number of observations in each total. Add these together and subtract the CF. This gives the SS(time).	5. $\mathrm{SS(time)} = \frac{t_1^2}{n_1} + \frac{t_2^2}{n_2} + \frac{t_3^2}{n_3} - \mathrm{CF}$ $\mathrm{SS(time)} = \frac{(26{\cdot}4)^2}{8} + \frac{(19{\cdot}6)^2}{8} + \frac{(8{\cdot}8)^2}{8} - 125{\cdot}13$ $= 19{\cdot}69$
6. Square the total (T) of the observations at each temperature, divide by the number of observations in each total.	6. $\mathrm{SS(temp)} = \frac{T_1^2}{n_1} + \frac{T_2^2}{n_2} - \mathrm{CF}$ $= \frac{(30{\cdot}8)^2}{12} + \frac{(24{\cdot}0)^2}{12}$

Add these together and subtract the CF. This gives the SS(temp).

$$-125{\cdot}13$$
$$= 127{\cdot}05 - 125{\cdot}13$$
$$= 1{\cdot}92$$

7. Square the total (c) at each catalyst concentration and divide by the number of observations in each total. Add these together and subtract the CF. This gives the SS(cat).

7. $$\text{SS(cat)} = \frac{c_1^2}{n_1} + \frac{c_2^2}{n_2} - \text{CF}$$
$$= \frac{(39{\cdot}6)^2}{12} + \frac{(15{\cdot}2)^2}{12} - 125{\cdot}13$$
$$= 24{\cdot}80$$

8. Square the totals for given combinations of time and temperature in Table 2, divide by the number of original observations making up each total, and add. Subtract from this the CF, and SS(time) and SS(temp). The result is SS(time × temp).

8. SS(time × temp)
$$= \frac{(14{\cdot}4)^2}{4} + \frac{(11{\cdot}0)^2}{4} + \dots + \frac{(3{\cdot}4)^2}{4} - 125{\cdot}13 - 19{\cdot}69 - 1{\cdot}92$$
$$= 0{\cdot}02$$

9. Square the totals for given combinations of time and catalyst concentration in Table 3, divide by the number of original observations making up each total, and add. Subtract from this the CF, and SS(time) and SS(cat). The result is SS(time × cat).

9. SS (time × cat)
$$= \frac{(18{\cdot}6)^2}{4} + \frac{(7{\cdot}8)^2}{4} + \dots + \frac{(1{\cdot}8)^2}{4} - 125{\cdot}13 - 19{\cdot}69 - 24{\cdot}80$$
$$- 171{\cdot}60 - 125{\cdot}13 - 19{\cdot}69 - 24{\cdot}80 = 1{\cdot}98$$

10. Square the totals for given combinations of temperature and catalyst concentration in Table 4, divide by the number of original observations making up each total, and add. Subtract from this the CF, and SS(temp) and SS(cat). The result is SS(temp × cat).

10. $$\text{SS(temp × cat)} = \frac{(21{\cdot}6)^2}{6} + \frac{(9{\cdot}2)^2}{6} + \frac{(18{\cdot}0)^2}{6} + \frac{(6{\cdot}0)^2}{6} - 125{\cdot}13 - 1{\cdot}92 - 24{\cdot}80$$
$$= 151{\cdot}85 - 125{\cdot}13 - 1{\cdot}92 - 24{\cdot}80 = 0{\cdot}00$$

11. Square the totals of Table 1, divide each total by the number of individuals making up that total, and add. Subtract this total from the SSI.

11. SS(replication)
$$= 174{\cdot}70 - \frac{(10{\cdot}2)^2}{2} - \frac{(8{\cdot}4)^2}{2} - \dots - \frac{(0{\cdot}4)^2}{2} = 174{\cdot}70 - 173{\cdot}80$$

This is the ERROR or SS(REPLICATION).

$= 0{\cdot}90$

12. Finally, subtract all of the calculated SS from the TOTAL SS. The result is the highest order interaction, or SS(time × temp × cat).

12. SS(time × temp × cat)

$$= 49{\cdot}75 - 19{\cdot}69 - 1{\cdot}92 - 24{\cdot}80 - 0{\cdot}02 - 1{\cdot}98 - 0{\cdot}00 - 0{\cdot}90$$

$$= 0{\cdot}26$$

13. Set up the analysis of variance table as shown. The number of degrees of freedom is one less than the level of each factor. For interactions, the number of degrees of freedom is the product of the degrees of freedom for each factor. For replication, the number of degrees of freedom is given by the number of pairs tested. The mean square is the sum of squares divided by the degrees of freedom.

13. *Analysis of variance*

Source	Sum of squares	Degrees of freedom	Mean square
Time	19·69	2	9·840
Temperature	1·92	1	1·920
Catalyst	24·80	1	24·800
Time × temp	0·02	2	0·010
Time × cat	1·98	2	0·990
Temp × cat	0·00	1	0·010
Time × temp × cat	0·26	2	0·130
Replication	0·90	12	0·075
Total	49·57	23	

14. The mean squares (MS) are compared with the error term by means of the F ratio, where

$$F = \frac{\text{MS(factor)}}{\text{MS(replication)}}$$

Tables are available of critical values of F for the degree of significance required; and for the degrees of freedom in the factor and replication mean squares.

If the F value obtained exceeds the value in the table, we may say that the effect is real (not due to chance).

14. Starting with the highest order interaction:

$$F_{(\text{time} \times \text{temp} \times \text{cat})} = 0{\cdot}130/0{\cdot}075 = 1{\cdot}74$$

In the F table, the critical value for stating that an effect is real, with a 95% chance of being correct, is, for 2 and 12 df,

$$F_{0{\cdot}05\,(2,\,12)} = 3{\cdot}9$$

Since the calculated F does not exceed the critical value, we are not able to say that a real time × temperature × catalyst interaction exists.

15. In reporting significant effects, we sometimes use one asterisk to denote the 95% level of significance, two for 99%, and three for 99·9%.

The replication mean square, 0·075 in this case, is a measure of error variance. The square root of this number is the standard deviation of experimental + test error if the experiment actually was repeated twice; of test error, only, if the two results represent two analyses on each experiment.

The 95% limits of the experimental method (or the test error depending on the design used) can be estimated by multiplying the estimated standard deviation by two, and adding and subtracting this from the experimental result.

15. For the other factors:

Factor	F	Critical F	Significance
Temp × cat	1	4·8	Not sig.
Time × cat	13·2	3·9	Sig.**
Time × temp	1	3·9	Not sig.
Catalyst	330·6	4·8	Sig.***
Temperature	25·6	4·8	Sig.***
Time	131·2	3·9	Sig.***

$$S = \sqrt{0{\cdot}075} = 0{\cdot}27$$

$$95\% \text{ limits} = \pm 0{\cdot}54$$

This means that, out of a long series of experiments run at a given set of conditions, we would expect the individual results to lie between ±0·54 around the average.

The significant factors are catalyst, time, temperature, and the time × catalyst interaction. Reference to the two-way tables will help interpret the relative magnitude of the effects. In the time × catalyst interaction, the effect of change in catalyst level depends on how long the reaction has been run.

Using the information obtained in the analysis of variance illustrated, we should be able to readily obtain answers to the questions outlined at the beginning of the chapter. Reference to the stated bibliography will furnish the details in specific cases.

REGRESSION

Another kind of data analysis, which has much broader application than analysis of variance is called regression. This method has the same mathematical basis as analysis of variance, but in most cases the calculations become very long and tedious. Without computers, regression methods would be very little used. Since the computers arrived on the scene, the applications of regression methods in chemical experimentation have exceeded the applications of analysis of variance. Draper and Smith [9] present a sound and practical approach to regression analysis.

While in analysis of variance our main goal was to attach significance to the variables being studied, in regression we are interested in significance and magnitude of the effects, and also in representing the effects in the form of an equation. In the example shown under analysis of variance, the following terms are defined:

y = % residual monomer

x_1 = time $\quad x_2$ = temperature

x_3 = catalyst concentration

Then the following questions might be asked:

1. How can the relationship between y and the x's be expressed in the form of an equation?
2. What are the coefficients of the x terms which will give the best fit to the observed values of y?
3. Are all the coefficients necessary? Significant?
4. How close can y be predicted in this and future experiments?
5. Is the equation equally valid for all values of the x's?
6. Are the independent variables really independent?

In the example shown, a good starting place might be to consider an equation of the form

$$y = c_0 + c_1x_1 + c_2x_2 + c_3x_3$$

and fit this equation to the data by the method of least squares.

Or, looking at the experimental design, we see three levels of x_1. This may lead to the question: 'Is there curvature in the relationship of y with x_1?' Then an equation of one of the following forms might be helpful:

$$y = c_0 + c_1x_1 + c_{11}x_1^2 + c_2x_2 + c_3x_3$$

$$y = c_0 + c_1x_1 + c_{11}e^{-0\cdot 1x_1} + c_2x_2 + c_3x_3$$

$$\ln y = c_0 + c_1x_1 + c_2\ln x_2 + c_3\ln x_3$$

Selection of one of these equations would depend on the expected behaviour of the system, preliminary observations, or simply the best fit to the observed data.

Again, if we are interested in whether the x variables act independently, support one another (synergism), or oppose one another (antagonism), a model containing interaction terms is in order.

$$y = c_0 + c_1x_1 + c_2x_2 + c_3x_3 + c_{11}x_1^2 + c_{12}x_1x_2 + c_{13}x_1x_3 + c_{23}x_2x_3$$

1. $y = 2{\cdot}28 - 1{\cdot}10(x_1 - 2) - 0{\cdot}57(x_2 - 1{\cdot}5) - 2{\cdot}03(x_3 - 1{\cdot}5) \pm 0{\cdot}83$ (95% limits)

2. $y = 2{\cdot}45 - 1{\cdot}0(x_1 - 2) - 0{\cdot}57(x_2 - 1{\cdot}5) - 2{\cdot}03(x_3 - 1{\cdot}5) - 0{\cdot}25(x_1 - 2)^2 + 0{\cdot}05(x_1 - 2)(x_2 - 1{\cdot}5) + + 0{\cdot}07(x_2 - 1{\cdot}5)(x_3 - 1{\cdot}5) \pm 0{\cdot}54$ (95% limits)

Note: the x values are measured from their midpoints, hence the parenthetical variables in the equations. This helps to prevent certain undesirable happenings in the computer.

Looking at just two of these possible expressions for the relationships, these equations are obtained. At the same time, the regression programme calculates a residual standard deviation, which is a measure of how far the actual values differ from those produced by the equation.

Equation 1 predicts the values of the response by considering only the purely linear effects of the variables. Equation 2 employs linear, quadratic, and cross-product terms to produce a better prediction of the response (narrower limits of variation).

An even more complex model is possible for this situation. It includes all possible linear, quadratic, and linear-quadratic cross-product terms.

$$y = c_0 + c_1x_1 + c_2x_2 + c_3x_3 + c_{11}x_1^2 + c_{12}x_1x_2 + c_{13}x_1x_3 + c_{23}x_2x_3 + c_{112}x_1^2x_2 + c_{113}x_1^2x_3 + c_{123}x_1x_2x_3 + c_{1123}x_1^2x_2x_3$$

These are only a few of the possible mathematical models which might be used in the multiple regression approach to data analysis. Since the reason for fitting an equation to the data is to give greater insight into what is going on, the simplest model which adequately represents the behaviour of the response is usually preferable. If a theoretical model is available from kinetics or other considerations, it will give the most meaningful equation.

A useful computer programme available as a supplement to regression is known as the *response surface plot.* This programme calculates the values of the independent variables which combine to give a specific value of the response, and it plots contours of the specific values on coordinates made up of pairs of the independent variables.

Chart I

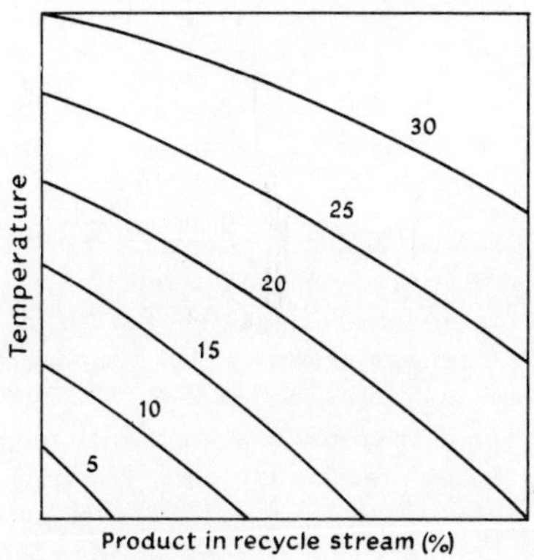

Chart I shows a plot of gross conversion of raw material as a function of temperature and recycle of product. The lines are contours of percentage of conversion and show that the increase in the percentage of conversion with percentage of product depends on the reaction temperature.

Chart II

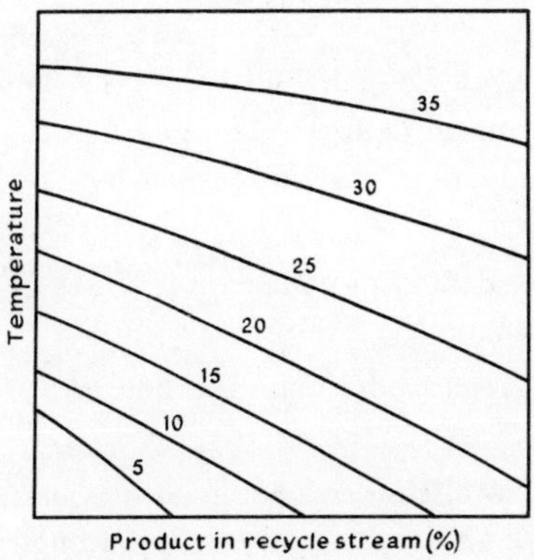

Chart II shows the same response at a higher level of activator. Comparison with Chart I shows that activator gives higher conversion at high temperatures but lower conversion at low temperatures.

Chart III

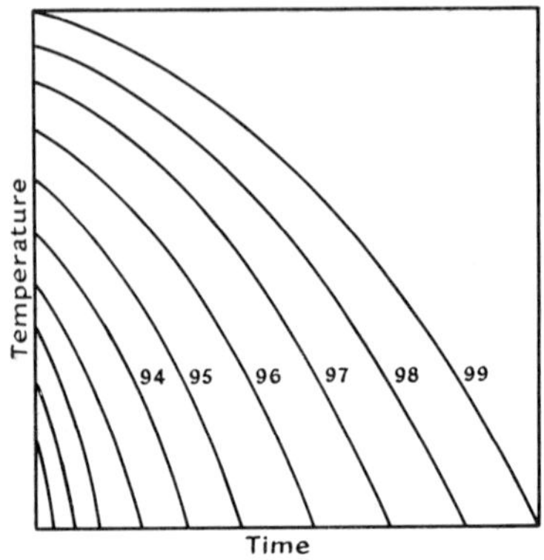

Chart III shows yield as a function of time and temperature. Note that yield increases rapidly with time and temperature at the lower levels; less rapidly as optimum yield is approached.

Chart IV

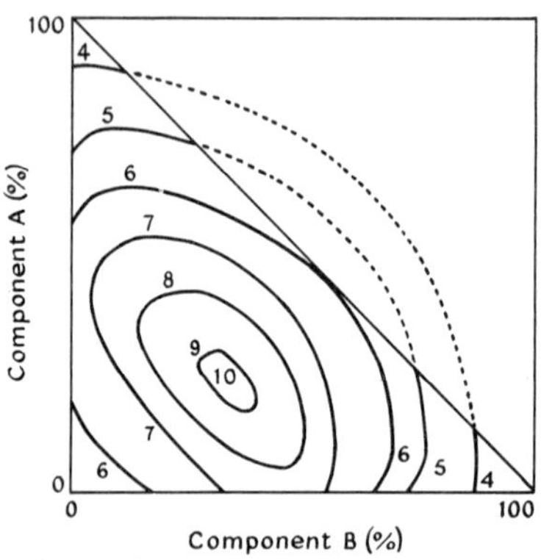

Chart IV shows oxidative stability of a three-component additive mixture. Contour peak shows that a maximum exists at 20% component A and 35% component B. The concentration of component C is found by subtracting the other two from 100. Dotted lines show calculated contours with concentrations of A and B adding up to more than 100%.

Other features of the regression programme permit us to determine:

1. How well balanced is the experimental design? The degree of relationship among pairs of 'independent variables' may be calculated.
2. How good is the prediction? Calculation of confidence limits of the predicted response at any point in the experimental space.
3. Is there a better equation? Any given model equation can be tested for 'lack of fit'.
4. Are we missing any variables? The difference between the actual and predicted values of the response can be analysed for trends with time, or for relationship with other variables not included in the equation.

5. Can the equation be simplified? Systematic inclusion or exclusion of a variable, or of terms of a given order of magnitude may indicate a simpler and more meaningful expression.
6. Can the same equational form be used to predict many responses? Several dependent variables can be included in one pass, and the coefficients of the equation calculated for each.
7. Are there relationships among the dependent or response variables? By rearranging the data, the correlations between all possible pairs may be evaluated.

In all applications of multiple regression which involve equations of more than three terms, a digital computer programme is practically a must. In using the analysis of variance, a fairly useful rule of thumb is that up to 100 data points is not too much to handle by the desk calculator route.

PROBLEMS

1. Solve problem 1, Chapter IV, by analysis of variance. Set significance level $\alpha = 0{\cdot}05$.

A	*B*
97·2	96·5
96·5	96·9
97·4	96·3
96·8	96·0
96·8	96·1
	96·2
484·7	578·0

$$n = 11$$

$$\mathrm{GT} = 1062{\cdot}7$$

$$\frac{(1062{\cdot}7)^2}{11} = 102666{\cdot}48$$

$$\text{SS individual} = 102668{\cdot}53 - 102666{\cdot}48 = 2{\cdot}05$$

$$\text{SS between sets} = (484{\cdot}7)^2/5 - (578{\cdot}0)^2/6 - 102666{\cdot}48$$

$$= 46986{\cdot}82 + 55680{\cdot}67 - 102666{\cdot}48$$

$$= 1{\cdot}01$$

$$\text{SS within sets} = 1{\cdot}04$$

Source	SS	df	MS	F
Between sets	1·01	1	1·01	
Within sets	1·04	9	0·116	8·71
Total	2·05	10		

From Table Ib, the critical value for $F_{1,9} = 5{\cdot}3$. We can say with a probability of 0·95 of being correct that there is a difference between the two sets of data. Not that the analysis of variance can be used for comparing the means of two sets of data, and will give the same conclusions as the t test.

2. Prior to the start of a research project, an experimenter wanted to know if three lots of a chemical could be used interchangeably in the experimental programme. The synthesis in question had been carried out using these different lots and yield results were as follows:

Lot I	*Lot II*	*Lot III*
96·2	96·4	96·8
96·1	96·1	97·2
97·2	95·7	96·9
96·5	288·2	97·1
386·0		97·0
		485·0

$$\text{Total} = 1159{\cdot}2$$
$$n = 12$$

Is the experimenter justified, at the 0·05 level of significance, in concluding that these materials do not give comparable results?

$$\text{SS individual} = 111981{\cdot}50 - 111978{\cdot}72 = 2{\cdot}78$$
$$\text{SS lots} = (386{\cdot}0)^2/4 + (288{\cdot}2)^2/3 + (485{\cdot}0)^2/5 - 111978{\cdot}72$$
$$= 37249{\cdot}00 + 27686{\cdot}41 + 47045{\cdot}00 - 111978{\cdot}72$$
$$= 1{\cdot}69$$
$$\text{SS within lots} = 2{\cdot}78 - 1{\cdot}69 = 1{\cdot}09$$

Source	SS	df	MS	F
Between lots	1·69	2	0·845	
Within lots	1·09	9	0·121	6·98
Total	2·78	11		

$$F_{0\cdot05;2,9\mathrm{df}} = 4\cdot3$$

Materials differ significantly in yield and should not be used interchangeably.

INSTANT STATISTICS

'What is all this about interaction?'

The inner circle in any activity is recognized by the facility with which the members use the jargon. It almost seems like an organized plot to keep the uninitiated from understanding what the others are talking about. Statisticians are no exception, and, when the secret word is one which might have meaning in other fields besides statistics, confusion reigns.

Take *interaction*, for example. Imagine a statistical report that says: 'in this experiment, time and temperature are not significant but the time × temperature *interaction* is.' Or 'This experimental design does not permit evaluation of second-order *interactions*.' What does this *interaction* mean?

Defining *interaction* is something like asking someone to describe a spiral staircase with his hands tied behind his back. The example is much more effective than the verbal description. Let us take an example therefore.

In an experiment set up to determine the effect of time and catalyst concentration on the yield of a reaction, the following results were obtained (results shown are averages of two determinations):

Time (min)	*Catalyst*	*Yield* (%)
20	0·1	85
40	0·1	88
60	0·1	90
20	0·2	87
40	0·2	90
60	0·2	87
20	0·3	91
40	0·3	92
60	0·3	85

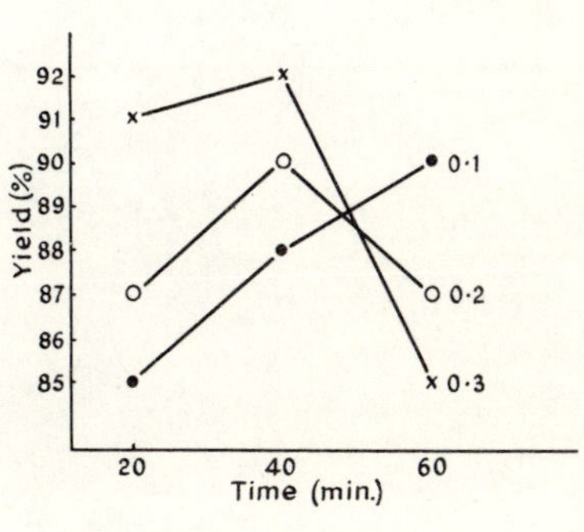

The statistical analysis comes back with the statement that time and catalyst concentration are not significant, but the interaction between time and catalyst concentration is significant at the 99·9% confidence level.

To use the verbal definition, interaction means that the effects of the factors on the response are not independent; that is, the effect of one factor on the response is not the same at each level of another factor. In the example given, increasing the catalyst concentration gave improved yields at 20 min and 40 min reactions, the opposite at 60 min reaction. Or, conversely, with 0·1% catalyst, yield increased with time up to 60 min; with 0·2% and 0·3% catalyst, yield went through a maximum at 40 min, and then decreased.

Evaluation and explanation of interactions by the chemist and statistician very often results in a better understanding of basic mechanisms.

Interactions can involve more than two factors. If we explain an interaction of $A \times B$ by saying that the effect of factor A is not the same at all levels of factor B, then an $A \times B \times C$ interaction simply means that the effect of A and B jointly is not the same at all levels of C. You can build up step-by-step to higher order interactions.

A final example:

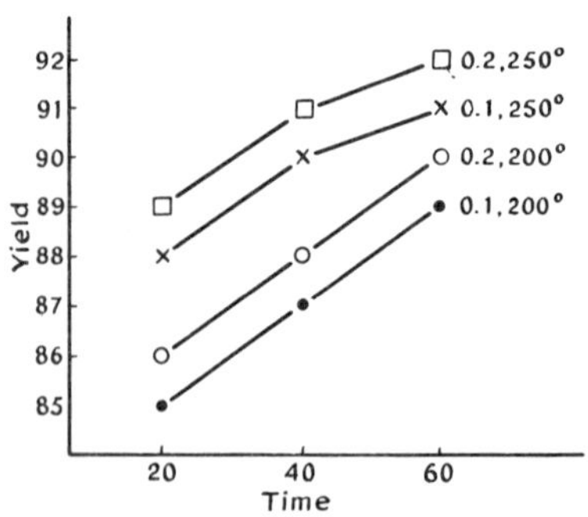

No interaction

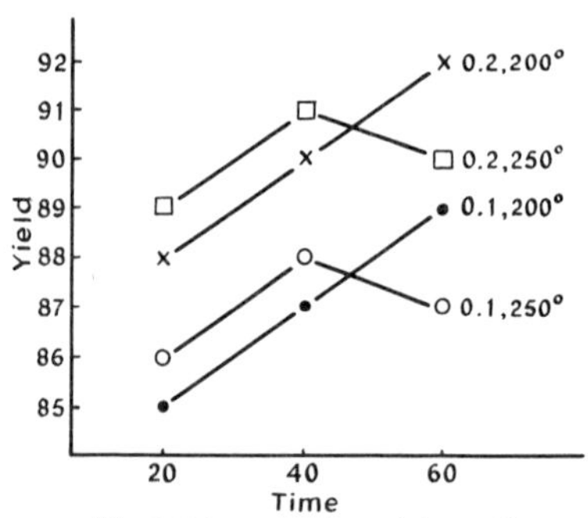

Time x temperature interaction

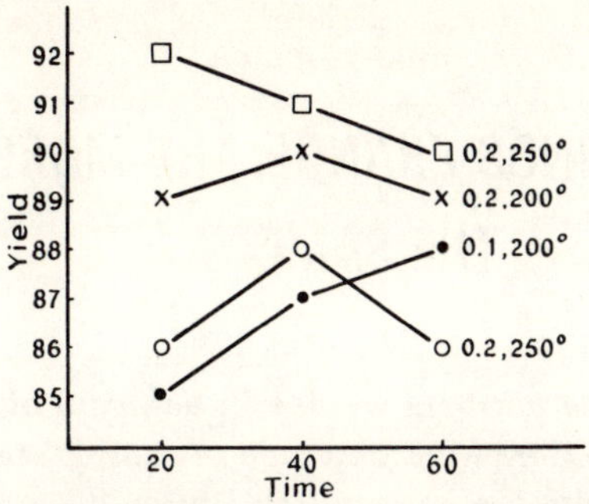

Time x temperature x catalyst interaction

The challenge in interpreting interactions is not in determining whether they exist or not; this is readily calculated. The real challenge is to the chemist, to explain what is happening to cause these unusual results.

CHAPTER EIGHT

THE MORE THINGS CHANGE, THE MORE THEY ARE THE SAME

The behaviour of the numbers we use in analysis of data depends pretty much on how they were obtained. A solids' test, for example, may be measured with the same reliability at 40% solids as at 60% solids. A moisture determination measured at 0·5% may be more or less reproducible than one measured at 10%, depending on whether one considers absolute or relative deviations. The relationship between viscosity and solids may be curved; plotted on semi-logarithmic graph paper, it becomes a straight line. Volume resistivity data, tensile strength, percentage elongation, penescope measurement of size on paper – these may all require methods of treatment which differ from those used on yield, concentration, temperature, etc.

To make data more readily understandable, the use of transformations is sometimes advisable. Transformation involves changing the scale of the variables (independent, dependent, or both), to accomplish one or more of the following purposes :

1. To conform to theoretical considerations.
2. To make a symmetrical or normal distribution out of skewed data.
3. To remove interactions.
4. To make the variation independent of where we are on the scale of measurement.
5. To make linear relationships out of non-linear.

In selecting a transformation, the more that is known about the theoretical behaviour of the variables, the better is the choice that we make. In reaction rate data, for example, we may use the reciprocal of the absolute temperature in place of the measured centigrade value; we may use the logarithm of the rate in place of the calculated value. Viscosity data are usually expressed as logarithms when any analysis is made. Many situations exist in which the transformation is selected on theoretical grounds.

Other situations arise where, in trying to work towards determining a theoretical basis for explaining observed facts, we may not know what sort of transformation, if any, will help. One approach is to consider the nature of the variation. If the standard deviation can be calculated at various average values, and plotted against the averages, the following rules of thumb may be used:

1. Standard deviation independent of the mean.	1. No transformation required.
2. Standard deviation proportional to the square root of the mean.	2. Use the square root of the measured value.
3. Standard deviation proportional to the mean.	3. Use the logarithm of the measured value.
4. Standard deviation proportional to the square of the mean.	4. Use the reciprocal of the measured value.
5. Standard deviation increases up to some middle value of the mean, then decreases. (Data expressed as a percentage or proportion.)	5. Use the angle (in radians or degrees) whose sine is the square root of the observed value. (If values of 0 and 1 are not actually obtainable, the logarithm of $x/(1-x)$, where x is the observed value lying between 0 and 1, may be used.)

The value of these transformations is that, on the transformed scale, the calculated standard deviation is independent of where we are on the scale. If this were not the case, we would tend to overestimate the variability for small values of y, and underestimate it for large values of y.

If there are not enough data to calculate meaningful standard deviations at various levels, plotting the variables in pairs may often lead to the proper transformation. The following charts show how curved lines can be straightened out by the proper choice of transformations. In these cases, the variation is usually made constant at the same time. The left column shows the untransformed relationship; the right shows the results of the transformation.

EFFECT OF SOME COMMON TRANSFORMATIONS

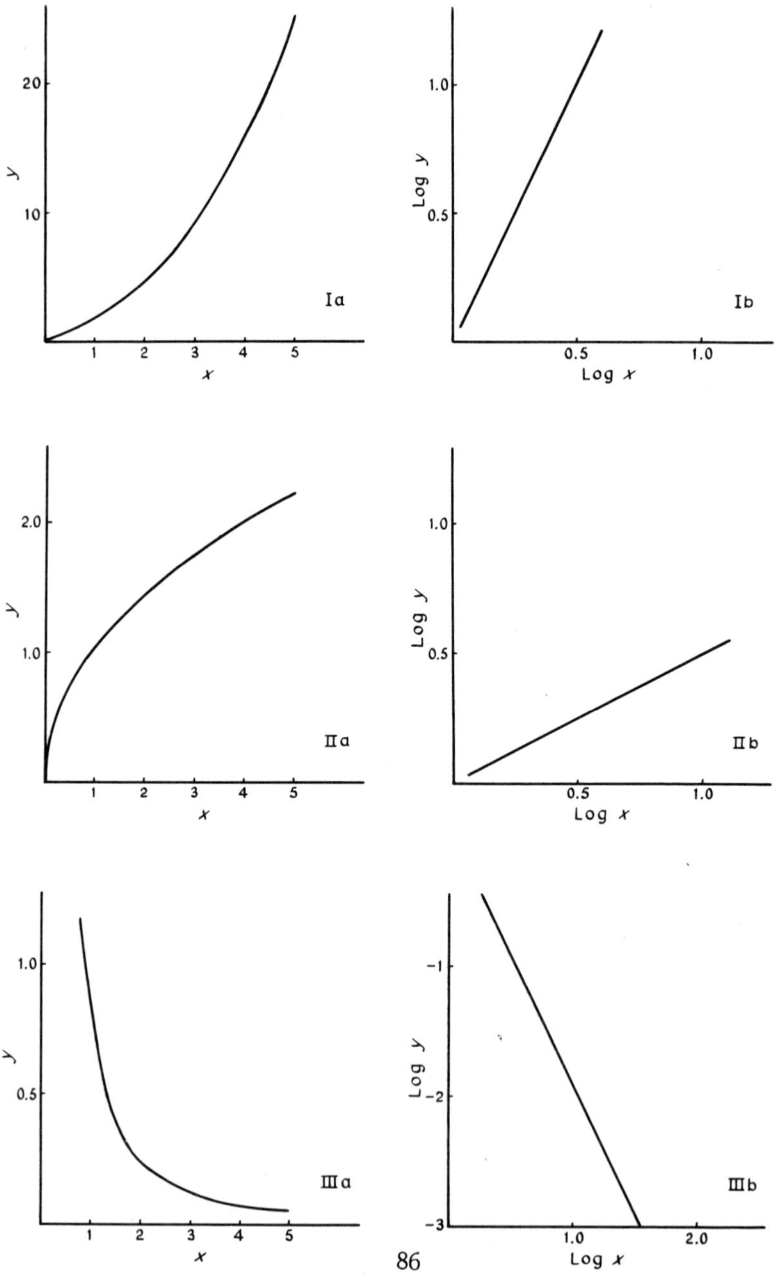

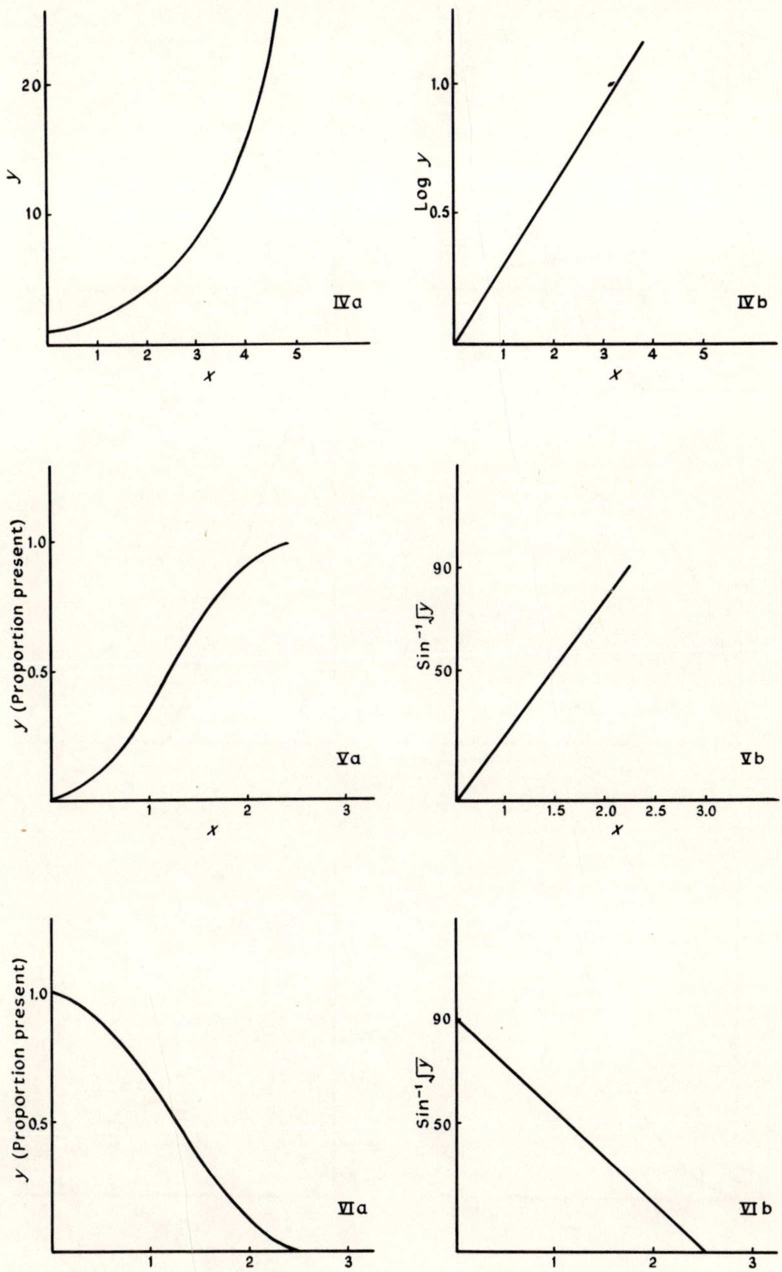
y
20
10
1
2
3
4
5
x
IVa
Log y
1.0
0.5
1
2
3
4
5
x
IVb
y (Proportion present)
1.0
0.5
1
2
3
x
Va
Sin⁻¹√y
90
50
1
1.5
2.0
2.5
3.0
x
Vb
y (Proportion present)
1.0
0.5
1
2
3
x
VIa
Sin⁻¹√y
90
50
1
2
3
x
VIb

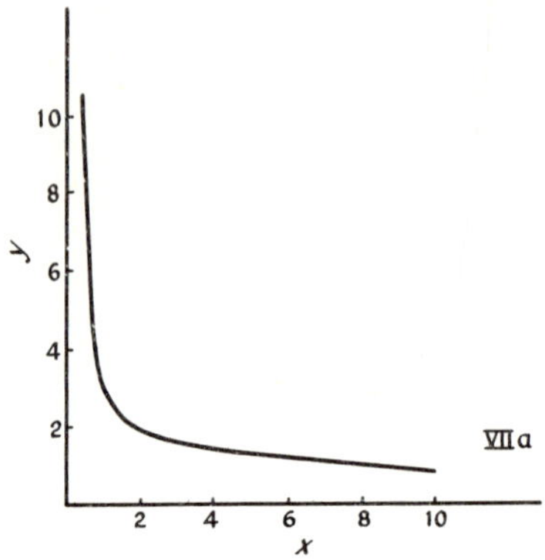
y
10
8
6
4
2
2
4
6
8
10
x
VII a

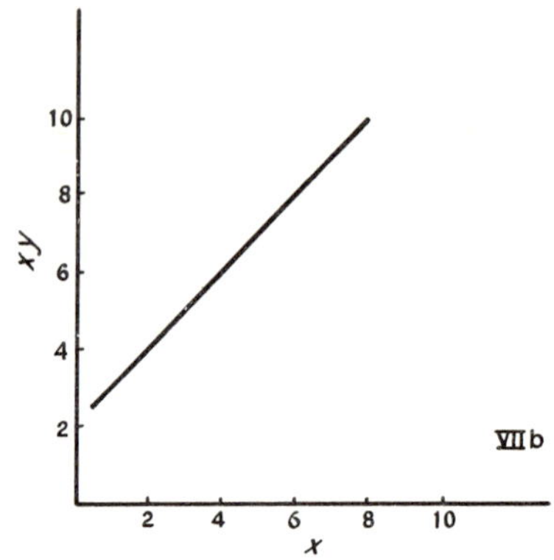
xy
10
8
6
4
2
2
4
6
8
10
x
VII b

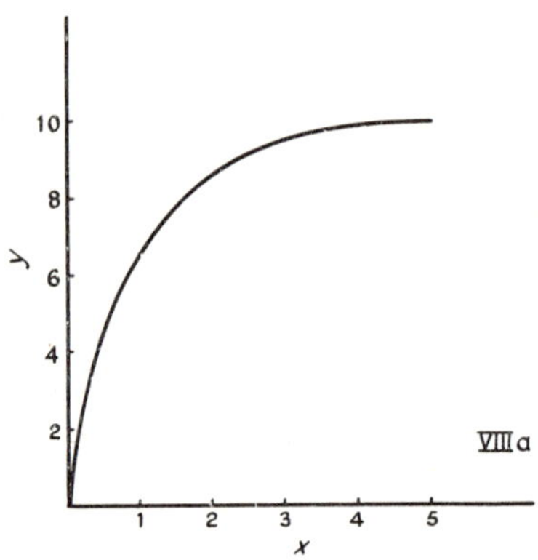
y
10
8
6
4
2
1
2
3
4
5
x
VIII a

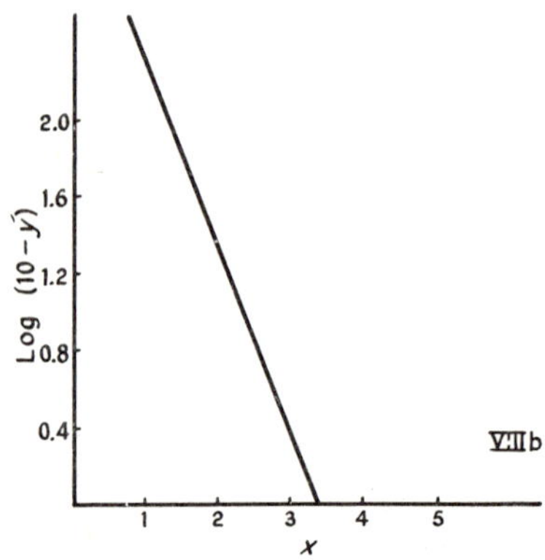
Log (10 − y)
2.0
1.6
1.2
0.8
0.4
1
2
3
4
5
x
V:II b

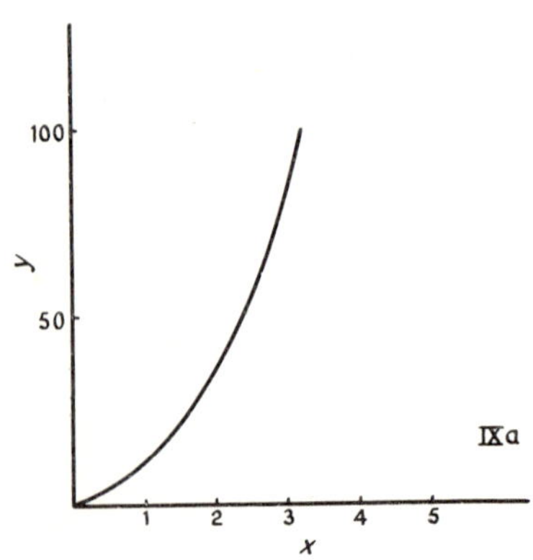
y
100
50
1
2
3
4
5
x
IX a

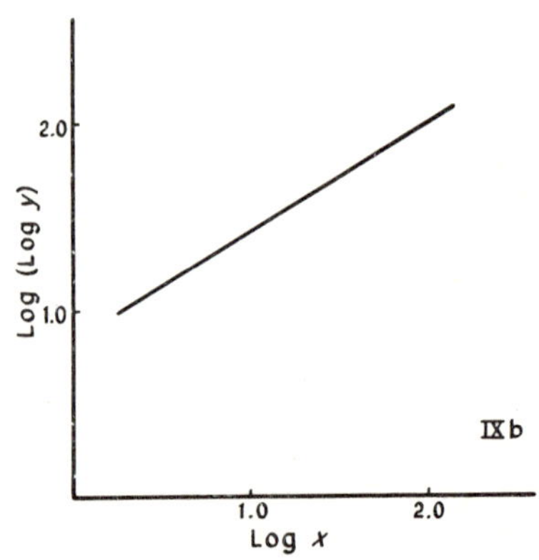
Log (Log y)
2.0
1.0
1.0
2.0
Log x
IX b

Chart no.	*Basic relationship*	*Transformation*
I	$y = ax^b$	$\log y = \log a + b \log x$
II	$y = ax^{1/b}$	$\log y = \log a + (\log x/b)$
III	$y = ax^{-b}$	$\log y = \log a - b \log x$
IV	$y = ab^x$	$\log y = \log a + x \log b$
V, VI	Increase or decrease of proportion with time $y = f(x)$	Replace y by $\sin^{-1}\sqrt{y}$ $\sin^{-1}\sqrt{y} = f'(x)$
VII	$y = a + (b/x)$	$xy = ax + b$
VIII	$y = a - be^{-x}$	$\ln(a - y) = -x + \ln b$
IX	$y = ae^{(bx)^n}$	$\ln(\log y/a) = n \ln b + n \ln x$

INSTANT STATISTICS

Watch your language in mixed company

'*Homoscedasticity* can be of real concern in fields other than abnormal psychology. Some *deviates* are known to be *homoscedastic*, and others are definitely *heteroscedastic*. Statisticians, in general, much prefer *homoscedastic deviates* to those other kind.'

A definition of terms is definitely needed. First of all, a deviate is any numerical value which is subject to variation. And *homoscedastic* means that the variation is independent of the absolute value of the deviate.

Most of the statistical tests we use (t test, F test, analysis of variance, multiple regression analysis) are predicated on the assumption that the variation being studied is the same, regardless of whether the property averages 10 or 50 or 75,000. For example, a homoscedastic variable might show variation as follows (several measurements on the same sample:)

Sample A	*Sample B*	*Sample C*
10	137	1492
11	138	1493
9	136	1491
11	138	1493
10	137	1492
9	136	1491

A heteroscedastic variable might show variation as :

Sample D	*Sample E*	*Sample F*
10	137	1492
11	151	1641
9	123	1343
11	151	1641
10	137	1492
9	123	1343

Within fairly narrow ranges of a measured variable (e.g. 5 – 15, 100 – 200, 1000 – 2000), the variable may be treated as is, with the assumption that the data are homoscedastic. Over a wider range, in order to get the most out of statistical analysis, some sort of transformation is required.

A good example of the effect of transformations on the variation of data around some given true value is found in viscosity measurements. Here the variability is definitely related to the level of viscosity. However, the logarithm of the viscosity is homoscedastic, as can be seen below.

	Viscosity	*95% limits*	*Range*	*Log viscosity*	*95% limits*	*Range*
1	10	8 – 12	4	1·00	0·90 – 1·08	0·18
2	60	48 – 72	24	1·78	1·68 – 1·86	0·18
3	350	280 – 420	140	2·54	2·45 – 2·63	0·18
4	1737	1389·6 – 2084·4	694·8	3·24	3·14 – 3·32	0·18

Transformations of data are also used to linearize relationships involving the variable in question. For example, the relationship between solids and viscosity of a resin solution may be sharply curved; the relationship between solids and log viscosity is a straight line.

If we keep our designs orthogonal and our data homoscedastic, our decisions will be uniform no matter which way we look at them.

GLOSSARY OF TERMS

Accuracy the closeness of an estimate or measurement to the exact or true value.

Alias in a partial factorial design, certain treatments cannot be distinguished. These are known as **aliases.**

Arithmetic mean (or Average) the sum of a set of values divided by the number of values.

Array an arrangement of a set of observations.

Attribute a qualitative characteristic of an individual material or object.

Autocorrelation relationship between successive observations in a series.

Average see arithmetic mean.

Average deviation (or Mean Deviation) the arithmetic mean of the absolute differences between a set of observed and expected observations.

Bernoulli distribution the Binomial Distribution.

Bias an effect which *systematically* distorts a statistical result.

Bimodal distribution a distribution having two peaks or **modes.**

Binomial distribution if an event has a probability p of occurring in one trial, the binomial distribution gives its probability of occurring r times in n trials. The parameter r has only discrete whole number values, and the value of p at each r is given by

$$p = \binom{n}{r} q^{n-r} p^r; \quad \binom{n}{r} = \text{the number of combinations of } n \text{ things taken } r \text{ at a time,}$$

$$q = 1 - p.$$

Block in experimental design, a group of items under treatment or observation.

Canonical variate in correlations, the value of an observation after transforming the direction and scale of the coordinate axes. **Canonical transformation** is the procedure by which this is done.

Central limit theorem if n independent variates have finite variances, their sum will tend to be normally distributed as n increases.

Chi square distribution the distribution of the sum of the squares of n independent normal variates in standard form. It is used for testing the deviation of observed from expected frequencies in counted data.

Confidence interval the interval within which a parameter being estimated may be said to lie, with some given degree of probability.

Confidence limits the limits of a confidence interval.

Confidence level the probability associated with a confidence interval.

Confidence region a multidimensional confidence interval.

Confounding a device whereby, in large factorial experiments, the number of runs to be made can be reduced by sacrificing some of the possible comparisons. The comparisons thus sacrificed are said to be confounded.

Continuous variate observations which are capable of assuming an infinite number of values along a continuous scale of measurement.

Correlation the interdependence of two or more variates.

Correlation coefficient a measure of the interdependence of two or more variates. In the case of two variates, a correlation coefficient of 1 indicates a perfect relationship in which the variates change in the same direction. A correlation coefficient of –1 indicates a perfect relationship in which the variates change in opposite directions. A correlation coefficient of zero denotes a complete absence of a systematic linear relationship.

Covariance analysis a technique for separating the treatment effects on a given variate from those due to relationship with another variate which in turn is related to the treatments.

Critical value the value of a **statistic** corresponding to a given significance level.

Degrees of freedom the number of values which can be assigned arbitrarily to a set within the specification of a system ; or the number of independent comparisons which can be made within the members of a sample.

Deviate the value of a variate measured from some standard reference point, usually the mean.

Dispersion the degree of scatter shown by observations.

Distribution free method a method for testing a hypothesis or setting up a confidence interval, which does not depend on the form of the underlying distribution.

Distribution function an equation which describes the relative frequency with which an observation of a given magnitude may be expected to occur.

Error the difference between an occurring value and its 'expected' or 'true' value.

Experimental error the variation to be expected under repetition of the experiment excluding mistakes in design or avoidable imperfections in technique.

Error of the first kind (α error or type I error) the probability of rejecting a true hypothesis.

Error of the second kind (β error or type II error) the probability of accepting a false hypothesis.

Error of the third kind the probability of giving the right answer to the wrong question (improper selection of hypothesis).

Estimator a rule or method for estimating a parameter of the parent population, usually from an incomplete sample.

Experimental design a plan or arrangement for obtaining data by experimental means. The basis for a particular experimental design will depend on the kind of information desired and the purpose of the experimental programme. Generally speaking, the design will permit unambiguous and efficient analysis of the data.

***F*-test** (The Variance Ratio Test) a method for determining, with a given degree of probability, whether the variances of two populations differ significantly from one another.

Factor in experimental design, a quantity under examination in an experiment as a possible cause of variation.

Factorial experiment an experiment designed to examine the effect of two or more factors, each applied at least at two levels of operation. The full factorial investigates all possible combinations of these factors at the indicated levels.

Fractional replication a factorial experiment in which only a balanced fraction of the possible treatment combinations is run.

Frequency the number of occurrences of a given event, or the number of members of a set falling into a particular class.

Frequency distribution a graphical or numerical presentation of the way in which the members of a set of observations are distributed.

Graeco-Latin square an experimental design which permits study of the effects of 4 factors at n levels in n^2 runs ($n \geqslant 4$).

Heteroscedastic having a variance which changes with the magnitude of the observations.

Histogram a frequency diagram in which rectangles proportional in area to the observed frequencies are created at intervals on the horizontal axis.

Homoscedastic having a variance which is independent of the magnitude of the observations.

Hypothesis (Statistical) a statistical hypothesis is an hypothesis concerning the parameter or form of a probabilistic mechanism which is assumed to generate the observations.

Independent variable in experimental design, the independent variables are the conditions which are deliberately controlled at chosen intervals. In regression analysis, if a response y is expressed as a function of x_1, x_2, etc., the x's are called the independent variables.

Inflation factor in regression analysis, a term which evaluates the interdependence of the independent variables. A low inflation factor is desirable for use of the regression equation for prediction.

Interaction a measure of the extent to which the effect (on the dependent variable) of one factor depends upon the level of another factor.

Invariance the degree to which a set of observations is unchanged by a particular transformation.

Level the value at which a factor is controlled during an experiment.

Linear constraint a condition imposed on certain variables which is linear in form. For example, in a three-component mixture, conc. x_1 + conc. x_2 + conc. $x_3 = 1$ is a linear constraint.

Linear correlation a coefficient of correlation constructed from linear functions of the observations.

Log normal distribution a distribution of observations which is normal when the logarithms of the measured data are used.

Main effect an estimate of the effect of an experimental variable measured independently of the other variables in the experiment.

Mean any one of a number of estimates of the centre of a set of observations (e.g. arithmetic mean, geometric mean, harmonic mean).
Arithmetic Mean – the average of the observed values.
Geometric Mean – the nth root of the product of n values.
Harmonic Mean – the reciprocal of the average of the reciprocals of the observed values.

Mean deviation the average of the absolute deviations of the observations from some central value.

Median the value of an observation which divides the total number of observations in half.

Mid-range the average of the two extreme values of a set of observations.

Mode the value in a set of observations which occurs most frequently – the 'peak' of a distribution curve.

Model a formalized expression of a theory, or of a causal situation which is assumed to have created the observed data. A mathematical model is such an expression in the form of an equation or set of equations.

Monte Carlo method a procedure for solving problems by constructing an artificial model representing a process and performing sampling experiments on it.

Moving average the average of n successive values in a time series, where a new average is formed by dropping the earliest value and adding the next observation.

Multiple correlation coefficient an index which measures the joint effect of several variables on some response.

Multiple regression the regression of a dependent variable on more than one independent variable.

Noise a convenient term for purely random variation.

Non-parametric methods statistical tests which make no assumptions about the distributions from which the data are obtained. These can be used to show differences, relationships, or association even when the characteristic observed can not be measured numerically.

Normal deviate the value of a deviate of the normal distribution.

Normal distribution a continuous frequency distribution in which the ordinate is given by

$$y = \left\{ \frac{1}{\sigma \sqrt{(2\pi)}} \right\} \left\{ \exp\left[-\frac{1}{2} \left(\frac{x - m}{\sigma} \right)^2 \right] \right\}$$

m = mean ; σ = standard deviation.

Normal equations a set of equations expressing the coefficients sought in regression analysis as the unknowns, and using functions calculated from the observed data as known constants.

Orthogonality in experimental design, the situation where the levels of the factors are balanced with respect to one another (statistical independence).

Parameter the unknown 'true' value of the basic properties defining a population or distribution.

Poisson distribution the exponential limit of a binomial distribution as $p \to 0$ and $n \to \infty$.

Population any finite or infinite collection of individuals.

Power the power of a statistical test is the probability that it will reject an alternate hypothesis when that hypothesis is false.

Precision a measure of the degree to which repeated observations of the same thing agree among themselves.

Random having a probability of occurrence determined by some probability distribution.

Range the difference between the largest and smallest value in a set of observations.

Regression a statistical method for investigating the relationship between a dependent variable or response, and one or more independent or predicated variables. If a response y can be expressed as a systematic function of x plus an error term (ϵ):

$$y = f(x) + \epsilon$$

then the systematic portion $y = f(x)$ is called the regression of y on x. This reasoning can be extended to many independent x terms, and the relationship can vary quite widely in complexity.

Regression coefficient the coefficient of an independent variable in a regression equation. If $y = c_1x_1 + c_2x_2$, c_1 and c_2 are the regression coefficients.

Replication repetition of an experiment in order to improve estimation and reduce experimental error.

Residual in regression analysis, the difference between an observed value and the value predicted by the regression equation; in analysis of variance, the error remaining after all desired main effects and interactions have been calculated.

Response the change in the dependent variable resulting from changes in the independent variable. Sometimes used to denote the dependent variable itself.

Response surface a multidimensional representation of the behaviour of the dependent variable as a function of one or more independent variables.

Response surface plot a graphical representation of the response surface as a contour map of the dependent variable on a coordinate scale of two of the independent variables.

Scatter diagram a diagram showing the joint variation of two variable quantities x and y.

Sign test a test of significance based on the signs of certain observations and not their magnitude.

Significant having an estimated probability of occurrence due to chance less than some predetermined probability level.

Simplex design an experimental design based on the minimum symmetrical geometric pattern possible. This pattern will have a dimension of one less than the number of variables. Hence, for 4 variables, the Simplex design is a tetrahedron, for 3 variables a triangle, for 2 variables a line, and for a single variable a point.

Simplex method a method for calculating the optimum allocation of scarce resources in a restricted system.

Standard deviation the square root of the average of the squares of the deviations of individual observations from their arithmetic mean.

Statistic a value calculated from a sample of observations and used to make some judgement about the observations. Sometimes used to estimate a parameter of a distribution.

Stochastic implying the presence of a random variable.

Student's *t* a statistic used for testing differences in level between two sets of data when the variability of the data is only estimate of the population variation

Student's *t* test a significance test using Student's *t* as a criterion.

Two-Tailed test a test for significance by which a hypothesis may be rejected when the test statistic is either too large or too small to permit acceptance of the hypothesis.

Variable a quantity which varies; a quantity which may take on any one of a specified set of values.

Variance the mean of the squares of the deviations of observations from their arithmetic mean.

Variance, Analysis of a method for partitioning the total variation of a set of observations into components due to specified factors. Comparison of these components with the portion of the variation due to experimental error is then made.

Variance ratio test a test used to determine the difference in variability between two sets of data. Used in the analysis of variance to compare variation due to a particular factor with the experimental error.

Variate a quantity which may take on any one of a specified set of values, with a specified probability.

BIBLIOGRAPHY

[1] Brownlee, K.A., *Industrial Experimentation,* Chemical Publishing Company.

[2] Brownlee, K.A., *Statistical Theory and Methodology in Science and Engineering,* Wiley.

[3] Cochran, W.G. and Cox, G.M., *Experimental Designs,* Wiley.

[4] Davies, O.L., *The Design and Analysis of Industrial Experiments,* Hafner.

[5] Dixon, W.J., and Massey, F.J., *Introduction to Statistical Analysis,* McGraw-Hill.

[6] Grant, E.L., *Statistical Quality Control,* McGraw-Hill.

[7] Olmstead, P.S., and Tukey, J.W., 'A Corner Test for Association', *Annals of Mathematical Statistics* **18**, 495–513 (1947).

[8] Quenouille, M.H., *Rapid Statistical Calculations,* Hafner.

[9] Draper, N.R., and Smith, H., *Applied Regression Analysis,* Wiley.

[10] Box, G.E.P., 'Evolutionary Operation: A Method for Increasing Industrial Productivity', *Applied Statistics* **VI**, No. 2 (1957).

[11] Box, G.E.P., and Hunter, J.S., 'Condensed Calculations for Evolutionary Operation Programs', *Technometrics* **1**, No.1 (Feb., 1959).

KEY TO USE OF TABLES

Table I (a, b, c, d) *Tables of the variance ratio*

Suppose one had an inexhaustible store of numbers which were normally distributed with a variance σ^2. Reaching into this distribution, say we take a random sample of $N_1 + 1$ of these numbers; then another random sample of $N_2 + 1$ of these numbers. If we calculate the variances of these samples, calling them S_1^2 and S_2^2, and then form the ratio:

$$F = S_1^2/S_2^2$$

the probability is 0·20 that F will be larger than the value in Table Ia corresponding to N_1 and N_2. For example, if we took 25 and 10 observations, we would expect the value of F to exceed 1·7 (corresponding to 24 and 9), 20% of the time. The value 1·7 is called the critical value of F for 24 and 9 degrees of feedom, for $\alpha = 0{\cdot}20$.

Tables Ib, Ic, and Id give the critical values of F for $\alpha = 0{\cdot}05$, 0·01, and 0·001, respectively.

For a one-tailed test, α will be equal to 0·20, 0·05, 0·01, and 0·001 in Tables Ia to Id. For a two-tailed test, α will be equal to twice these values.

In the analysis of variance, the comparisons are made using the α values for the one-tailed situation. In comparing two observed variances, the one-tailed test is used when we are asking whether the population variance represented by S_1^2 is larger than that represented by S_2^2; two-tailed test when we are asking, 'Are they equal?'

Table II *Substitute F ratio, ratio of ranges*

This table represents the upper and lower tails of the distribution of the ratios of ranges of samples taken from a normal distribution of variance σ^2. If the ratio R_1/R_2 of the ranges R_1 and R_2, with size N_1 and N_2 respectively, is calculated, the probability that it will be less than the value in the table corresponding to N_1 and N_2 is given in the column headed cum. prop. For example, if the distribution is normal and the ratio R_1/R_2 for 5 and 8 observations is calculated, the probability that the ratio will be less than 1·8 is 0·95.

For one-tailed tests on observed data from unknown sources, the critical values in the table are for $(1 - \alpha)$. For two-tailed tests, the critical values are for $\alpha/2$ and $1-(\alpha/2)$ for the low and high ends of the distribution.

Table III *Table of the t distribution*

For the one-tailed test, the table is to be read from the top, with the column headings representing the value of $1 = \alpha$.

For the two-tailed test, the columns are read from top and bottom, with the column headings representing $1 - (1/2\alpha)$ and $1/2\alpha$ respectively.

For example, if two means are being compared, and we want to limit the error of the first kind to $\alpha = 0{\cdot}05$, and we have 15 degrees of freedom in the data.

For a one-tailed test, we say the means are different if t exceeds 1·753. For a two-tailed test, we say the means are different if t exceeds 2·131 or is less than −2·131.

Table IV *Substitute t ratios*

This table is used for comparing an observed mean with some standard value, using the range as a measure of variability. The criteria for decision are the same as those for Table III.

Table V *Substitute t ratios*

This table is used to compare two observed means, using the average of their ranges as a measure of variability. The criteria for decision are the same as those for Table III.

Table VI *Table of the correlation coefficient*

If random samples are drawn in pairs from a normal distribution, and the correlation coefficient calculated as described in Chapter V, the probability that the absolute value of r will exceed the value in the table for the corresponding degrees of freedom is given at the top of each column. The degrees of freedom are 2 less than the number of pairs (n) used in the calculation.

Supposing the correlation coefficient is calculated from two sets of data from unknown sources. If the calculated value of r exceeds the value in the table, we make the statement that the two sets of data are linearly related. The probability at the top of the column represents our chance of being wrong in making such a statement.

TABLES*

Table Ia

Tables of variance ratio (i)

N_2 \ N_1	0·20 significance level								
	1	2	3	4	5	6	12	24	∞
1	9·5	12·0	13·1	13·7	14·0	14·3	14·9	15·2	15·6
2	3·6	4·0	4·2	4·2	4·3	4·3	4·4	4·4	4·5
3	2·7	2·9	2·9	3·0	3·0	3·0	3·0	3·0	3·0
4	2·4	2·5	2·5	2·5	2·5	2·5	2·5	2·4	2·4
5	2·2	2·3	2·3	2·2	2·2	2·2	2·2	2·2	2·1
6	2·1	2·1	2·1	2·1	2·1	2·1	2·0	2·0	2·0
7	2·0	2·0	2·0	2·0	2·0	2·0	1·9	1·9	1·8
8	2·0	2·0	2·0	1·9	1·9	1·9	1·8	1·8	1·7
9	1·9	1·9	1·9	1·9	1·9	1·8	1·8	1·7	1·7
10	1·9	1·9	1·9	1·8	1·8	1·8	1·7	1·7	1·6
11	1·9	1·9	1·8	1·8	1·8	1·8	1·7	1·6	1·6
12	1·8	1·8	1·8	1·8	1·7	1·7	1·7	1·6	1·5
13	1·8	1·8	1·8	1·8	1·7	1·7	1·6	1·6	1·5
14	1·8	1·8	1·8	1·7	1·7	1·7	1·6	1·6	1·5
15	1·8	1·8	1·8	1·7	1·7	1·7	1·6	1·5	1·5
16	1·8	1·8	1·7	1·7	1·7	1·6	1·6	1·5	1·4
17	1·8	1·8	1·7	1·7	1·7	1·6	1·6	1·5	1·4
18	1·8	1·8	1·7	1·7	1·6	1·6	1·5	1·5	1·4
19	1·8	1·8	1·7	1·7	1·6	1·6	1·5	1·5	1·4
20	1·8	1·8	1·7	1·7	1·6	1·6	1·5	1·5	1·4
22	1·8	1·7	1·7	1·6	1·6	1·6	1·5	1·4	1·4
24	1·7	1·7	1·7	1·6	1·6	1·6	1·5	1·4	1·3
26	1·7	1·7	1·7	1·6	1·6	1·6	1·5	1·4	1·3
28	1·7	1·7	1·7	1·6	1·6	1·6	1·5	1·4	1·3
30	1·7	1·7	1·6	1·6	1·6	1·5	1·5	1·4	1·3
40	1·7	1·7	1·6	1·6	1·5	1·5	1·4	1·4	1·2
60	1·7	1·7	1·6	1·6	1·5	1·5	1·4	1·3	1·2
120	1·7	1·6	1·6	1·5	1·5	1·5	1·4	1·3	1·1
∞	1·6	1·6	1·6	1·5	1·5	1·4	1·3	1·2	1·0

See previous section for key to use of tables.

Table Ib

Table of variance ratio (ii)

N_2 \ N_1	*0·05 significance level*								
	1	2	3	4	5	6	12	24	∞
1	164·4	199·5	215·7	224·6	230·2	234·0	234·9	249·0	254·3
2	18·5	19·2	19·2	19·3	19·3	19·3	19·4	19·5	19·5
3	10·1	9·6	9·3	9·1	9·0	8·9	8·7	8·6	8·5
4	7·7	6·9	6·6	6·4	6·3	6·2	5·9	5·8	5·6
5	6·6	5·8	5·4	5·2	5·1	5·0	4·7	4·5	4·4
6	6·0	5·1	4·8	4·5	4·4	4·3	4·0	3·8	3·7
7	5·6	4·7	4·4	4·1	4·0	3·9	3·6	3·4	3·2
8	5·3	4·5	4·1	3·8	3·7	3·6	3·3	3·1	2·9
9	5·1	4·3	3·9	3·6	3·5	3·4	3·1	2·9	2·7
10	5·0	4·1	3·7	3·5	3·3	3·2	2·9	2·7	2·5
11	4·8	4·0	3·6	3·4	3·2	3·1	2·8	2·6	2·4
12	4·8	3·9	3·5	3·3	3·1	3·0	2·7	2·5	2·3
13	4·7	3·8	3·4	3·2	3·0	2·9	2·6	2·4	2·2
14	4·6	3·7	3·3	3·1	3·0	2·9	2·5	2·3	2·1
15	4·5	3·7	3·3	3·1	2·9	2·8	2·5	2·3	2·1
16	4·5	3·6	3·2	3·0	2·9	2·7	2·4	2·2	2·0
17	4·5	3·6	3·2	3·0	2·8	2·7	2·4	2·2	2·0
18	4·4	3·6	3·2	2·9	2·8	2·7	2·3	2·1	1·9
19	4·4	3·5	3·1	2·9	2·7	2·6	2·3	2·1	1·9
20	4·4	3·5	3·1	2·9	2·7	2·6	2·3	2·1	1·8
22	4·3	3·4	3·1	2·8	2·7	2·6	2·2	2·0	1·8
24	4·3	3·4	3·0	2·8	2·6	2·5	2·2	2·0	1·7
26	4·2	3·4	3·0	2·7	2·6	2·5	2·2	2·0	1·7
28	4·2	3·3	3·0	2·7	2·6	2·4	2·1	1·9	1·7
30	4·2	3·3	2·9	2·7	2·5	2·4	2·1	1·9	1·6
40	4·1	3·2	2·9	2·6	2·5	2·3	2·0	1·8	1·5
60	4·0	3·2	2·8	2·5	2·4	2·3	2·9	1·7	1·4
120	3·9	3·1	2·7	2·5	2·3	2·2	1·8	1·6	1·3
∞	3·8	3·0	2·6	2·4	2·2	2·1	1·8	1·5	1·0

Table Ic

Tables of variance ratio (iii)

N_2 \ N_1	*0·01 significance level*									
	1	2	3	4	5	6	8	12	24	∞
1	4052	4999	5403	5625	5764	5859	5981	6106	6234	6366
2	98·5	99·0	99·2	99·3	99·3	99·4	99·3	99·4	99·5	99·5
3	34·1	30·8	29·5	28·7	28·2	27·9	27·5	27·1	26·6	26·1
4	21·2	18·0	16·7	16·0	15·5	15·2	14·8	14·4	13·9	13·5
5	16·3	13·3	12·1	11·4	11·0	10·7	10·3	9·9	9·5	9·0
6	13·7	10·9	9·8	9·2	8·8	8·5	8·1	7·7	7·3	6·9
7	12·3	9·6	8·5	7·9	7·5	7·2	6·8	6·5	6·1	5·7
8	11·3	8·7	7·6	7·0	6·6	6·4	6·0	5·7	5·3	4·9
9	10·6	8·0	7·0	6·4	6·1	5·8	5·5	5·1	4·7	4·3
10	10·0	7·6	6·6	6·0	5·6	5·4	5·1	4·7	4·3	3·9
11	9·7	7·2	6·2	5·7	5·3	5·1	4·7	4·4	4·0	3·6
12	9·3	6·9	6·0	5·4	5·1	4·8	4·5	4·2	3·8	3·4
13	9·1	6·7	5·7	5·2	4·9	4·6	4·3	4·0	3·6	3·2
14	8·9	6·5	5·6	5·0	4·7	4·5	4·1	3·8	3·4	3·0
15	8·7	6·4	5·4	4·9	4·6	4·3	4·0	3·7	3·3	2·9
16	8·5	6·2	5·3	4·8	4·4	4·2	3·9	3·6	3·2	2·8
17	8·4	6·1	5·2	4·7	4·3	4·1	3·8	3·5	3·1	2·7
18	8·3	6·0	5·1	4·6	4·3	4·0	3·7	3·4	3·0	2·6
19	8·2	5·9	5·0	4·5	4·2	3·9	3·6	3·3	2·9	2·5
20	8·1	5·9	4·9	4·4	4·1	3·9	3·6	3·2	2·9	2·4
22	7·9	5·7	4·8	4·3	4·0	3·8	3·5	3·1	2·8	2·3
24	7·8	5·6	4·7	4·2	3·9	3·7	3·3	3·0	2·7	2·2
26	7·7	5·5	4·6	4·1	3·8	3·6	3·3	3·0	2·6	2·1
28	7·6	5·5	4·6	4·1	3·8	3·5	3·2	2·9	2·5	2·1
30	7·6	5·4	4·5	4·0	3·7	3·5	3·2	2·8	2·5	2·0
40	7·3	5·2	4·3	3·8	3·5	3·3	3·0	2·7	2·3	1·8
60	7·1	5·0	4·1	3·7	3·3	3·1	2·8	2·5	2·1	1·6
120	6·9	4·8	4·0	3·5	3·2	3·0	2·7	2·3	2·0	1·4
∞	6·6	4·6	3·8	3·3	3·0	2·8	2·5	2·2	1·8	1·0

Table Id

Table of variance ratio (iv)

N_2 \ N_1	0·001 significance level									
	1	2	3	4	5	6	8	12	24	∞
1	varying from 400,000 to 600,000									
2	998	999	999	999	999	999	999	999	999	999
3	167	148	141	137	135	133	131	128	126	123
4	74·1	61·3	56·2	53·4	51·7	50·5	49·0	47·4	45·8	44·1
5	47·0	36·6	33·2	31·1	29·8	28·8	27·6	26·4	25·1	23·8
6	35·5	27·0	23·7	21·9	20·8	20·0	19·0	18·0	16·9	15·8
7	29·2	21·7	18·8	17·2	16·2	15·5	14·6	13·7	12·7	11·7
8	25·4	18·5	15·8	14·4	13·5	12·9	12·0	11·2	10·3	9·3
9	22·9	16·4	13·9	12·6	11·7	11·1	10·4	9·6	8·7	7·8
10	21·0	14·9	12·6	11·3	10·5	9·9	9·2	8·5	7·6	6·8
11	19·7	13·8	11·6	10·4	9·6	9·1	8·3	7·6	6·9	6·0
12	18·6	13·0	10·8	9·6	8·9	8·4	7·7	7·0	6·3	5·4
13	17·8	12·3	10·2	9·1	8·4	7·9	7·2	6·5	5·8	5·0
14	17·1	11·8	9·7	8·6	7·9	7·4	6·8	6·1	5·4	4·6
15	16·6	11·3	9·3	8·3	7·6	7·1	6·5	5·8	5·1	4·3
16	16·1	11·0	9·0	7·9	7·3	6·8	6·2	5·6	4·9	4·1
17	15·7	10·7	8·7	7·7	7·0	6·6	6·0	5·3	4·6	3·9
18	15·4	10·4	8·5	7·5	6·8	6·4	5·8	5·1	4·5	3·7
19	15·1	10·2	8·3	7·3	6·6	6·2	5·6	5·0	4·3	3·5
20	14·8	10·0	8·1	7·1	6·5	6·0	5·4	4·8	4·2	3·4
22	14·4	9·6	7·8	6·8	6·2	5·8	5·2	4·6	3·9	3·2
24	14·0	9·3	7·6	6·6	6·0	5·6	5·0	4·4	3·7	3·0
26	13·7	9·1	7·4	6·4	5·8	5·4	4·8	4·2	3·6	2·8
28	13·5	8·9	7·2	6·3	5·7	5·2	4·7	4·1	3·5	2·7
30	13·3	8·8	7·1	6·1	5·5	5·1	4·6	4·0	3·4	2·6
40	12·6	8·2	6·6	5·7	5·1	4·7	4·2	3·6	3·0	2·2
60	12·0	7·8	6·2	5·3	4·8	4·4	3·9	3·3	2·7	1·9
120	11·4	7·3	5·8	5·0	4·4	4·0	3·5	3·0	2·4	1·6
∞	10·8	6·9	5·4	5·6	4·1	3·7	3·3	2·7	2·1	1·0

Table II

Substitute F ratio, ratio of ranges

Sample size for denominator	*Cum. prop.*	*Sample size for numerator*								
		2	3	4	5	6	7	8	9	10
2	0·005	0·0078	0·096	0·21	0·30	0·38	0·44	0·49	0·54	0·57
	0·01	0·0157	0·136	0·26	0·38	0·46	0·53	0·59	0·64	0·68
	0·025	0·039	0·217	0·37	0·50	0·60	0·68	0·74	0·79	0·83
	0·05	0·079	0·31	0·50	0·62	0·74	0·80	0·86	0·91	0·95
	0·95	12·7	19·1	23	26	29	30	32	34	35
	0·975	25·5	38·2	52	57	60	62	64	67	68
	0·99	63·7	95	116	132	142	153	160	168	174
	0·995	127	191	230	250	260	270	280	290	290
3	0·005	0·0052	0·071	0·16	0·24	0·32	0·38	0·43	0·47	0·50
	0·01	0·0105	0·100	0·20	0·30	0·37	0·43	0·49	0·53	0·57
	0·025	0·026	0·160	0·28	0·39	0·47	0·54	0·59	0·64	0·68
	0·05	0·052	0·23	0·37	0·49	0·57	0·64	0·70	0·75	0·80
	0·95	3·19	4·4	5·0	5·7	6·2	6·6	6·9	7·2	7·4
	0·975	4·61	6·3	7·3	8·0	8·7	9·3	9·8	10·2	10·5
	0·99	7·37	10	12	13	14	15	15	16	17
	0·995	10·4	14	17	18	20	21	22	23	25
4	0·005	0·0043	0·059	0·14	0·22	0·28	0·34	0·39	0·43	0·46
	0·01	0·0086	0·084	0·18	0·26	0·33	0·39	0·44	0·48	0·52
	0·025	0·019	0·137	0·25	0·34	0·42	0·48	0·53	0·57	0·61
	0·05	0·043	0·20	0·32	0·42	0·50	0·57	0·62	0·67	0·70
	0·95	2·02	2·7	3·1	3·4	3·6	3·8	4·0	4·2	4·4
	0·975	2·27	3·5	4·0	4·4	4·7	5·0	5·2	5·4	5·6
	0·99	3·83	5·0	5·5	6·0	6·4	6·7	7·0	7·2	7·5
	0·995	4·85	6·1	7·0	7·6	8·1	8·5	8·8	9·3	9·6
5	0·005	0·0039	0·054	0·13	0·20	0·26	0·32	0·36	0·40	0·44
	0·01	0·0076	0·079	0·17	0·24	0·31	0·36	0·41	0·45	0·49
	0·025	0·018	0·124	0·23	0·32	0·38	0·44	0·49	0·53	0·57
	0·05	0·038	0·18	0·29	0·40	0·46	0·52	0·57	0·61	0·65
	0·95	1·61	2·1	2·4	2·6	2·8	2·9	3·0	3·1	3·2
	0·975	2·01	2·6	2·9	3·2	3·4	3·6	3·7	3·8	3·9
	0·99	2·64	3·4	3·8	4·1	4·3	4·6	4·7	4·9	5·0
	0·995	3·36	4·1	4·6	4·9	5·2	5·5	5·7	5·9	6·1

Sample size for denominator	Cum. prop.	Sample size for numerator 2	3	4	5	6	7	8	9	10
6	0·005	0·0038	0·051	0·12	0·19	0·25	0·30	0·35	0·38	0·42
	0·01	0·0070	0·073	0·16	0·23	0·29	0·34	0·39	0·43	0·46
	0·025	0·017	0·115	0·21	0·30	0·36	0·42	0·46	0·50	0·54
	0·05	0·035	0·16	0·27	0·36	0·43	0·49	0·54	0·58	0·61
	0·95	1·36	1·8	2·0	2·2	2·3	2·4	2·5	2·6	2·7
	0·975	1·67	2·1	2·4	2·6	2·8	2·9	3·0	3·1	3·2
	0·99	2·16	2·7	3·0	3·2	3·4	3·6	3·7	3·8	3·9
	0·995	2·67	3·1	3·5	3·8	4·0	4·1	4·3	4·5	4·6
7	0·005	0·0037	0·048	0·12	0·18	0·24	0·29	0·33	0·37	0·40
	0·01	0·0066	0·069	0·15	0·22	0·28	0·33	0·37	0·41	0·45
	0·025	0·016	0·107	0·20	0·28	0·34	0·40	0·44	0·48	0·52
	0·05	0·032	0·15	0·26	0·35	0·41	0·47	0·51	0·55	0·59
	0·95	1·26	1·6	1·8	1·9	2·0	2·1	2·2	2·3	2·4
	0·975	1·48	1·9	2·1	2·3	2·4	2·5	2·6	2·7	2·8
	0·99	1·87	2·3	2·6	2·8	2·9	3·0	3·1	3·2	3·3
	0·995	2·28	2·7	2·9	3·1	3·3	3·5	3·6	3·7	3·8
8	0·005	0·0036	0·045	0·11	0·18	0·23	0·28	0·32	0·36	0·39
	0·01	0·0063	0·065	0·14	0·21	0·27	0·32	0·36	0·40	0·43
	0·025	0·016	0·102	0·19	0·27	0·33	0·38	0·43	0·47	0·50
	0·05	0·031	0·14	0·25	0·33	0·40	0·45	0·50	0·53	0·57
	0·95	1·17	1·4	1·6	1·8	1·9	1·9	2·0	2·1	2·1
	0·975	1·36	1·7	1·9	2·0	2·2	2·3	2·3	2·4	2·5
	0·99	1·69	2·1	2·3	2·4	2·6	2·7	2·8	2·8	2·9
	0·995	2·03	2·3	2·6	2·7	2·9	3·0	3·1	3·2	3·3
9	0·005	0·0035	0·042	0·11	0·17	0·22	0·27	0·31	0·35	0·38
	0·01	0·0060	0·062	0·14	0·21	0·26	0·31	0·35	0·39	0·42
	0·025	0·015	0·098	0·18	0·26	0·32	0·37	0·42	0·46	0·49
	0·05	0·030	0·14	0·24	0·32	0·38	0·44	0·48	0·52	0·55
	0·95	1·10	1·3	1·5	1·6	1·7	1·8	1·9	1·9	2·0
	0·975	1·27	1·6	1·8	1·9	2·0	2·1	2·1	2·2	2·3
	0·99	1·56	1·9	2·1	2·2	2·3	2·4	2·5	2·6	2·6
	0·995	1·87	2·1	2·3	2·5	2·6	2·7	2·8	2·9	3·0
10	0·005	0·0034	0·041	0·10	0·16	0·22	0·26	0·30	0·34	0·37
	0·01	0·0058	0·060	0·13	0·20	0·26	0·30	0·34	0·38	0·41
	0·025	0·015	0·095	0·18	0·25	0·31	0·36	0·41	0·44	0·48
	0·05	0·029	0·13	0·23	0·31	0·37	0·43	0·47	0·51	0·54
	0·95	1·05	1·3	1·4	1·5	1·6	1·7	1·8	1·8	1·9
	0·975	1·21	1·5	1·6	1·8	1·9	1·9	2·0	2·0	2·1
	0·99	1·47	1·8	1·9	2·1	2·2	2·2	2·3	2·4	2·4
	0·995	1·75	2·0	2·2	2·3	2·4	2·5	2·6	2·6	2·7

Table III

Table of the t distribution

df	$t_{0\cdot60}$	$t_{0\cdot70}$	$t_{0\cdot80}$	$t_{0\cdot90}$	$t_{0\cdot95}$	$t_{0\cdot975}$	$t_{0\cdot99}$	$t_{0\cdot995}$
1	0·325	0·727	1·376	3·078	6·314	12·706	31·821	63·657
2	0·289	0·617	1·061	1·886	2·920	4·303	6·965	9·925
3	0·277	0·584	0·978	1·638	2·353	3·182	4·541	5·841
4	0·271	0·569	0·941	1·533	2·132	2·776	3·747	4·604
5	0·267	0·559	0·920	1·476	2·015	2·571	3·365	4·032
6	0·265	0·553	0·906	1·440	1·943	2·447	3·143	3·707
7	0·263	0·549	0·896	1·415	1·895	2·365	2·998	3·499
8	0·262	0·546	0·889	1·397	1·860	2·306	2·896	3·355
9	0·261	0·543	0·883	1·383	1·833	2·262	2·821	3·250
10	0·260	0·542	0·879	1·372	1·812	2·228	2·764	3·169
11	0·260	0·540	0·876	1·363	1·796	2·201	2·718	3·106
12	0·259	0·539	0·873	1·356	1·782	2·179	2·681	3·055
13	0·259	0·538	0·870	1·350	1·771	2·160	2·650	3·012
14	0·258	0·537	0·868	1·345	1·761	2·145	2·624	2·977
15	0·258	0·536	0·866	1·341	1·753	2·131	2·602	2·947
16	0·258	0·535	0·865	1·337	1·746	2·120	2·583	2·921
17	0·257	0·534	0·863	1·333	1·740	2·110	2·567	2·808
18	0·257	0·534	0·862	1·330	1·734	2·101	2·552	2·878
19	0·257	0·533	0·861	1·328	1·729	2·093	2·539	2·861
20	0·257	0·533	0·860	1·325	1·725	2·086	2·528	2·845
21	0·257	0·532	0·859	1·323	1·721	2·080	2·518	2·831
22	0·256	0·532	0·858	1·321	1·717	2·074	2·508	2·819
23	0·256	0·532	0·858	1·319	1·714	2·069	2·500	2·807
24	0·256	0·531	0·857	1·318	1·711	2·064	2·492	2·797
25	0·256	0·531	0·856	1·316	1·708	2·060	2·485	2·787
26	0·256	0·531	0·856	1·315	1·706	2·056	2·479	2·779
27	0·256	0·531	0·855	1·314	1·703	2·052	2·473	2·771
28	0·256	0·530	0·855	1·313	1·701	2·048	2·467	2·763
29	0·256	0·530	0·854	1·311	1·699	2·045	2·462	2·756
30	0·256	0·530	0·854	1·310	1·697	2·042	2·457	2·750
40	0·255	0·529	0·851	1·303	1·684	2·021	2·423	2·704
60	0·254	0·527	0·848	1·296	1·671	2·000	2·390	2·660
120	0·254	0·526	0·845	1·289	1·658	1·980	2·358	2·617
∞	0·253	0·524	0·842	1·282	1·645	1·960	2·326	2·576
df	$-t_{0\cdot40}$	$-t_{0\cdot30}$	$-t_{0\cdot20}$	$-t_{0\cdot10}$	$-t_{0\cdot05}$	$-t_{0\cdot025}$	$-t_{0\cdot01}$	$-t_{0\cdot005}$

When the table is read from the foot, the tabled values are to be prefixed with a negative sign. Interpolation should be performed using the reciprocals of the degrees of freedom.

Table IV

Substitute t ratios

$$\text{Percentiles* for } \tau_1 = \frac{\bar{X} - \mu}{w}$$

Sample size	P_{95}	$P_{97\cdot5}$	P_{99}	$P_{99\cdot5}$	$P_{99\cdot9}$	$P_{99\cdot95}$
2	3·175	6·353	15·910	31·828	159·16	318·31
3	0·885	1·304	2·111	3·008	6·77	9·58
4	0·529	0·717	1·023	1·316	2·29	2·85
5	0·388	0·507	0·685	0·843	1·32	1·58
6	0·312	0·399	0·523	0·628	0·92	1·07
7	0·263	0·333	0·429	0·507	0·71	0·82
8	0·230	0·288	0·366	0·429	0·59	0·67
9	0·205	0·255	0·322	0·374	0·50	0·57
10	0·186	0·230	0·288	0·333	0·44	0·50
11	0·170	0·210	0·262	0·302	0·40	0·44
12	0·158	0·194	0·241	0·277	0·36	0·40
13	0·147	0·181	0·224	0·256	0·33	0·37
14	0·138	0·170	0·209	0·239	0·31	0·34
15	0·131	0·160	0·197	0·224	0·29	0·32
16	0·124	0·151	0·186	0·212	0·27	0·30
17	0·118	0·144	0·177	0·201	0·26	0·28
18	0·113	0·137	0·168	0·191	0·24	0·26
19	0·108	0·131	0·161	0·182	0·23	0·25
20	0·104	0·126	0·154	0·175	0·22	0·24
	$-P_{05}$	$-P_{02\cdot5}$	$-P_{01}$	$-P_{0\cdot5}$	$-P_{0\cdot1}$	$-P_{0\cdot05}$

* When the table is read from the foot, the tabled values are to be prefixed with a negative sign.

Table V

Substitute t ratios

Percentiles* for $\tau_d = \dfrac{\bar{X}_1 - \bar{X}_2}{\frac{1}{2}(w_1 + w_2)}$

Sample sizes $N_1 = N_2$	P_{95}	$P_{97\cdot5}$	P_{99}	$P_{99\cdot5}$	$P_{99\cdot9}$	$P_{99\cdot95}$
2	2·322	3·427	5·553	7·916	17·81	25·23
3	0·974	1·272	1·715	2·093	3·27	4·18
4	0·644	0·813	1·047	1·237	1·74	1·99
5	0·493	0·613	0·772	0·896	1·21	1·35
6	0·405	0·499	0·621	0·714	0·94	1·03
7	0·347	0·426	0·525	0·600	0·77	0·85
8	0·306	0·373	0·459	0·521	0·67	0·73
9	0·275	0·334	0·409	0·464	0·59	0·64
10	0·250	0·304	0·371	0·419	0·53	0·58
11	0·233	0·280	0·340	0·384	0·48	0·52
12	0·214	0·260	0·315	0·355	0·44	0·48
13	0·201	0·243	0·294	0·331	0·41	0·45
14	0·189	0·228	0·276	0·311	0·39	0·42
15	0·179	0·216	0·261	0·293	0·36	0·39
16	0·170	0·205	0·247	0·278	0·34	0·37
17	0·162	0·195	0·236	0·264	0·33	0·35
18	0·155	0·187	0·225	0·252	0·31	0·34
19	0·149	0·179	0·216	0·242	0·30	0·32
20	0·143	0·172	0·207	0·232	0·29	0·31
	$-P_{05}$	$-P_{02\cdot5}$	$-P_{01}$	$-P_{0\cdot5}$	$-P_{0\cdot1}$	$-P_{0\cdot05}$

* When the table is read from the foot, the tabled values are to be prefixed with a negative sign.

Table VI

Table of the correlation coefficient

Degrees of freedom $(m-2)$	*r*				
	0·10	0·05	0·02	0·01	0·001
1	0·988	0·997	0·999	1·000	1·000
2	0·900	0·950	0·980	0·990	0·999
3	0·805	0·878	0·934	0·959	0·992
4	0·729	0·811	0·882	0·917	0·974
5	0·669	0·754	0·833	0·874	0·951
6	0·621	0·707	0·789	0·834	0·925
7	0·582	0·666	0·750	0·798	0·898
8	0·549	0·632	0·716	0·765	0·872
9	0·521	0·602	0·685	0·735	0·847
10	0·497	0·576	0·658	0·708	0·823
11	0·476	0·553	0·634	0·684	0·801
12	0·457	0·532	0·612	0·661	0·780
13	0·441	0·514	0·592	0·641	0·760
14	0·426	0·497	0·574	0·623	0·742
15	0·412	0·482	0·558	0·606	0·725
16	0·400	0·468	0·543	0·590	0·708
17	0·389	0·456	0·528	0·575	0·693
18	0·378	0·444	0·516	0·561	0·679
19	0·369	0·433	0·503	0·549	0·665
20	0·360	0·423	0·492	0·537	0·652
25	0·323	0·381	0·445	0·487	0·597
30	0·296	0·349	0·409	0·449	0·554
35	0·275	0·325	0·381	0·418	0·519
40	0·257	0·304	0·358	0·393	0·490
45	0·243	0·287	0·338	0·372	0·465
50	0·231	0·273	0·322	0·354	0·443
60	0·211	0·250	0·295	0·325	0·408
70	0·195	0·232	0·274	0·302	0·380
80	0·183	0·217	0·256	0·283	0·357
90	0·173	0·205	0·242	0·267	0·337
100	0·164	0·195	0·230	0·254	0·321

INDEX

Analysis of variance, 68,74
Assumptions in statistical testing, 14

Bayesian statistics, 66,67

Confidence limits of mean, 18,19
Correlation analysis, 33–46
Correlation coefficient, 35,101
Correlation by quick method, 38,46
Correlation with two variables, graphic method, 40,41,42

Definitions of statistical terms, 91 ff
Differences, Test for
- between sample mean and standard, 17,19
- between two means, 20,21
- between two means with differing variances, 25
- between paired sets of data, 23,24
- between variance of two samples, 12

Error, 4,5,6,7
Error of first kind, 14
Error of second kind, 14
Error, Measurement of, 7,8
Evolutionary operation, 64
Experimental designs, 48–63
- central composite designs, 52,53,54
- factorial designs, 49,50
- fractional factorial designs, 50,51
- Latin square designs, 51,52
- simplex designs, 56,57,58

Experimental designs for specific problems, 61,62,63
- determination of test error, 61
- effect of conditions on level and variability, 61
- establishment of optimum process conditions, 62
- optimum composition of a mixture, 63
- screening for existence of effects, 62
- screening ingredients in pairs, 63

Fitting straight lines to data, 36,37,39

Interaction, 49,50,73,81,82,83

Least significant difference, 22
Least squares line, 36,37
Linear programming, 65

Mean deviation, 8
Multiple regression analysis, 75,76

Outliers, elimination of, 26,27

Quadrant sum test, 38

Quick statistical methods,
- fitting straight line to data, 39
- substitute t test, 19,21,102
- substitute F test, 13,30,101
- tests for correlation, 38,46

Range as estimate of standard deviation, 9
Range ratio test, 13
Range, use in substitute t tests, 19,21
Rapid corner test for association, 38
Response surface plotting, 77,78

Standard deviation, 9
range estimate of, 9
Standard deviation of an average, 10
Statistical inference, 14
Student's t tests, 17,18,19,20,21, 102
Substitute t tests, 19,21,102

Transformations of scale of measurement, 84–89

Variables, controllable, uncontrollable, unknown, 5,6
Variance, comparison of two, 12
Variance, definition, 12
Variance ratio, 12,101
Variation, 4,5,6